D0711719

Thinking Critically to Solve Problems: Combining Values and College Mathematics

By Jacci and Scott White

Contents

Introduction

Benedictine heritage embraces the values of:
- Excellence
- Community
- Respect
- Personal development
- Responsible stewardship
- Integrity.

But what does that really mean? What does it mean for Mathematics? We hope this brief supplement to your text will help you to answer those questions.

Exercise 1 – Values

Take some time to write your own personal definitions of the six values listed above. Once you have your definitions, break into small groups (or online teams) and share your definitions. As a group, you are to construct a single definition for each value. Having done so, come up with one example of how a college student might experience/apply each value.

Discussion 1 (for class or small groups)

Consider the mission statement of Saint Leo University along with its educational & learning goals listed below.

Saint Leo University is a Catholic, liberal arts-based university serving people of all faiths.
Rooted in the 1,500-year-old Benedictine tradition, the university seeks balanced growth in mind, body, and spirit for all members of its community. At University Campus, at education centers, and through the Center for Online Learning, Saint Leo University offers a practical, effective model for life and leadership in a challenging world; a model based on a steadfast moral consciousness that recognizes the dignity, value, and gifts of all people.

Saint Leo University Mission and Values retrieved 3/22/15 from http://online.saintleo.edu/about-us/mission-values.aspx

Educational and Learning Goals

1. We expect students to demonstrate intellectual growth:
   - Think critically and independently
   - Make informed decisions
   - Commit to life-long learning
   - Engage in problem-solving
   - Exercise reasoned judgment
   - Develop quantitative skills
   - Learn experientially
   - Understand how living things and physical systems operate
   - Prepare for graduate study
2. We expect students to demonstrate effective communication skills:
   - Speak thoughtfully and respectfully
   - Listen carefully
   - Read critically

- Write clearly
- Present information well

3. We expect students to demonstrate deepened spiritual values:
   - Understand Catholic and Benedictine values and traditions
   - Commit to act in concert with one's values
   - Respect differences in belief systems and values
   - Show compassion and empathy
   - Understand the relationships among humans, living things, the universe and God
   - Balance one's life

4. We expect students to respond aesthetically:
   - Appreciate the beauty and balance in nature
   - Develop creativity
   - Demonstrate sensitivity
   - Visualize creative potential

5. We expect students to prepare for an occupation:
   - Strive for excellence
   - Develop an international perspective
   - Become competent in: managing people/tasks, responding to change, planning innovation, collaborating, applying technology, acting fiscally responsible

6. We expect students to demonstrate social responsibility:
   - Act with integrity
   - Exercise personal responsibility
   - Respect all living things
   - Work for diversity both locally and globally
   - Build community
   - Commit to resource stewardship

7. We expect students to demonstrate personal growth and development:

- Develop self-understanding
- Learn to manage self
- Deal with ambiguity
- Exercise flexibility
- Strengthen confidence and self-esteem
- Learn persistence
- Care for self and physical and spiritual well-being
- Develop leadership
- Foster a work ethic

8. We expect students to demonstrate effective interpersonal skills:
   - Value successful relationships
   - Participate effectively in group work
   - Cooperate
   - Engage in philanthropy
   - Volunteer

Writing Across the Curriculum using Critical Thinking and Values:

1. How can the mission and educational goals affect your approach to this course?

2. How do you feel the mission and educational goals can affect the behavior of your Math instructor?

3. How might the mission and educational goals make this Mathematics course different from its counterpart at another university?

4. What does the value of integrity have to do with your role as a student in this class? The role of your instructor?

5. Which of the other values should be a part of your experience in this class? How so?

6. How might the value of Community be the same and/or different between an online course and a face-to-face class?

Exercise 2

1. Choose one or more educational goals that you expect to achieve in this mathematics class and write an explanation of how you expect to achieve that goal.

2. Write a description of a personal goal that you aim to achieve in this Mathematics class.

3. Write a paragraph describing the educational goal you feel is the most important for this level of mathematics and explain why?

Exercise 3

1. Please review a newspaper and find an article that is meaningful to your childhood.

2. Use the article to introduce yourself to the class.

Chapter 1: Critical Thinking

Isn't Problem solving what mathematics is all about? You might be surprised to learn that a large part of problem solving stems from patterns. Patterns can be identified, providing information and the ability to predict. Or, patterns are often used in solutions for similar types of problems, or a familiar proof pattern might work unexpectedly on a fresh type of application.

Number Patterns

Often in number patterns, you look for a common arithmetic function to get you from one number to the next. For example, if you add the same number each time, do you get the next number in line? How about if you subtract, multiply, or divide by the same number? To test this strategy for the next number in line, if the numbers are increasing you try to determine how much you would add to the first number to get the second. If you add it again, will you get third and fourth... If not, then is there a number you can multiply the first one by to get the second. Will that common multiple allow you to get the next number in line? If the numbers decrease, then you might be looking for a common number to either subtract or divide in order to get the next number in line.

Example:

What is the next number in the pattern?
- a. 2, 5, 8, 11, ...
- b. 2, 6, 18, 54, ...
- c. 3/4, 3/8, 3/16, 3/32, ...
- d. 2, -1, -4, -7, ...

Solution:

- a. Add 3 to get 14.
- b. Multiply by 3 to get 162.
- c. Divide by 2 (or multiply by ½) to get 3/64.
- d. Subtract 3 to get -10.

Sometimes there can be an alternating pattern, so that every other number follows a pattern and you alternate between the two patterns. If the numbers are fractions, you can sometimes identify the pattern of the numerator and denominator separately, but the pattern would still be adding, subtracting, multiplying or dividing the prior numerator or denominator by something. Sometimes, rather than adding or subtracting the same number, the number might increase by one or two each time so you are adding 3, then 4, then 5, ... An alternating pattern could use two different arithmetic functions such as double a number, then add three, then double it, then add three, then double it, then add three, ...

If the sign in your pattern alternates between positive and negative, then you are probably multiplying or dividing by a negative number since those operations alternate signs. If the numbers are getting further from zero in the negative and positive direction, then it is probably multiplying by a negative. If they are alternating and getting closer to zero and even becoming fractions that alternate signs, the pattern is probably dividing by a negative number.

Patterns can also be with shapes or combinations of numbers. You might have ordered pairs such as points for a graph. If there are ordered pairs, it might be that there is no pattern to get from one pair of numbers to the next, but that there is a pattern to get from the first number in the pair to the second number in the pair, so that as long as you have the first number, you can get the second number in the pair. With patterns, sometimes they are straight forward, (literally) while other times they not only change shape, but also rotate or move in some way.

Inductive and Deductive Reasoning

Inductive and deductive reasoning also rely on patterns. Induction starts with observations such as patterns and works toward a generalization. The prior examples we discussed, of finding the pattern in order to continue it, would be inductive thinking. You see the numbers and test examples of how to get from one number to the next until you find a method that will work for all the numbers in the pattern and then use that same calculation to predict the next number in the sequence. If you have a pattern such as:

$$O \, \Phi \, O \, \Phi \, O \, \Phi$$

You can probably guess that a circle is the next shape. Or, if you are not sure if you can use a calculator on a test, so you look around and see someone using a calculator and the teacher does not stop them, then you determine that calculators are allowed on tests. The answer is implied but not explicitly stated.

Deductive thinking is when you start with a broad idea and work toward collecting examples in order to eventually make a conclusion about a specific situation. All the facts are there and they lead you to a specific and singular conclusion. For the calculator example, you might think you may use a calculator, so you pull out your syllabus to read the directions and see that it says no calculators allowed. You now deduce that you may not use a calculator for this test.

More broadly, inductive reasoning is when you come to a conclusion based on observations and experiences. Deductive reasoning is when you make a conclusion based on rules or facts. We have a lot of rules in mathematics. For example, adding a number can be considered a rule. Sometimes a mathematical rule can be made up of multiple steps so that you add a number, then divide by another number and finish by subtracting the number you started with.

Example:

Choose a number, now double it, add 4, divide the result in half and subtract your original number.

Solution:

Let x be the number

2*x	Double it
2*x+4	Add 4
$\dfrac{2x+4}{2} = x+2$	Divide by 2
x + 2 − x = 2	Subtract original number

You are left with the number 2.

Deductive reasoning was used with those rules and a general case of x to know that 2 will always be left over.

However, you could repeat this experiment by doing it with several different people.

Examples:

Let x = 3	Let x = 8	Let x = 0	Let x = -1
2*3 = 6	2*8 = 16	2*0 = 0	2*(-1) = -2
6+4 = 10	16+4 = 20	0+4 = 4	-2+4 = 2

10/2 = 5	20/2 = 10	4/2 = 2	2/2 = 1
5-3 = 2	10-8 = 2	2-0 = 2	1-(-1) = 2

Notice that you always end up with 2 so that you are using inductive reasoning to go from individual examples to the general conclusion that the result will always be the number 2.

Problem Solving

With problem solving, the most important first step is to read the problem enough times so you understand what the problem is about, and what you are looking for. Often, when someone struggles with a problem scenario, they are unable to tell what it is they are solving for within the problem.

Once you have read the problem several times and understand what it is about and what you are looking for, then you can usually let the unknown item be represented by x so when you solve for x, you have the solution to the problem. If there is more than one unknown quantity, then let one item be x and write the other unknowns in terms of x. For example, if you invest a total of $200 in two accounts, then the amount in one account might be x while the amount in the second account can them be written as 200-x rather than as a new variable.

You should now be able to take the information in the problem and relate it in an equation. This then allows you to solve the equation for the missing quantity.

Always check your solution in terms of the context of the original problem to make sure it makes sense.

Of course, not every problem fits this so easily. There may be times when a specific strategy is needed rather than this general strategy. In a case where you get stuck with the general strategy, you might consider drawing a picture or diagram. Often a visual image can make the problem clearer. Sometimes a table of possibilities can help you to identify the correct one. Or, you might recall a similar problem from the past and try whatever worked for you then. Sometimes all you can do is start with a guess, but often you can then use that guess to work your way toward the correct result.

Example:

A roving smoothie stand sold 27 medium smoothies for $3.50 each and 15 large smoothies that were $4.50 each. Expenses for the smoothies came to $37, how much profit was made on the smoothies?

Solution:

First read the problem several times. The question is asking for the profit. If we call the profit P, then an equation related to the profit would be the total income minus the costs. The income is the price of each smoothie times the number sold. The cost is given as $37. An equation expressing these quantities is:

P = income – costs

P = 27*3.5 + 15*4.5 – 37
P = 98 + 67.5 – 37
P = $128.50

Chapter Review

Number Patterns – a common arithmetic function to proceed from one number to the next in a series.

Inductive Reasoning – starts with observations such as patterns and works toward a generalization

Deductive Reasoning - when you make a conclusion based on rules or facts.

Explorations

In the News: Published Problems

Find a newspaper article that publishes a problem and the solution. The problem does not have to use numbers. Illustrate the steps that were used to solve the problem.

Critical Thinking:

In order to get the full amusement and understanding of mathematical jokes, you must fully understand the concepts behind the joke. Often, you not only need to understand one mathematical concept, but also a double meaning for the mathematical words that are used in the joke.

1. Find a math joke. Provide the joke as well as an explanation of the mathematics behind the joke. Do not forget to reference your source.

2. Mathematical riddles and games are a popular way to get young people to practice mathematics for fun. Find a mathematical riddle or game and explain the strategy and/or solution.

Values Discussion:

1. Many people find jokes to be humorous when the joke makes fun of a certain type of person. Math jokes are a good example of this. Find a joke that derives humor from putting down a mathematician. Do you believe the humor is offensive, funny, or

both? Explain why. How do these jokes affect the community?

2. Can interpreting math jokes improve your mathematics skills? How or why not?

3. How can jokes and riddles enhance community?

Critical Thinking with Values:

Use Inductive reasoning from what you have heard/seen/observed to make a conclusion about one of the following statements as true or false. Then, use the University Catalog or other resources to prove a conclusion using Deductive Reasoning.

1. Freshman cannot have cars on campus.

2. Every student is required to take a math course in order to graduate.

3. All students have a computer.

4. The school has a mission statement.

5. Every student must have credit in English in order to graduate.

Discussion:

1. Do you believe the result you discovered is fair? Support your response.

2. Are all colleges and universities the same with regards to each of the situations above? Which do you think are most important when choosing a college or university? Why?

3. How is the topic of freshman having cars on campus related to community, respect, and/or responsible stewardship?

4. How is the topic of all students having a computer related to the values of respect, community, and/or excellence?

5. Should all students be given a computer? If so, how should the cost be covered?

6. Should all students be required to complete the same basic requirements for a degree, or should some requirements be waived for individual students who might struggle in those areas?

7. How might some of these rules affect community?

Writing Across the Curriculum – Values, Critical Thinking, and Social Justice:

1. Describe an effective problem solving strategy and include examples of how it can be used.

2. Write a report explaining how mathematical problem solving can lead to excellence, personal development, and community outside of a mathematics classroom.

3. Document a problem from outside mathematics and illustrate how mathematical problem solving

can be used to solve the problem by finding an acceptable solution step by step.

4. How can effective problem solving be used to help someone or something less fortunate than you?

Chapter 2: Measurement

We measure many things, in many ways, in many places. It might be a length, weight, or volume. Or, it might be a percentage or portion of something. It might be converted between Metric and English measurement systems, within one system or the other, or between decimals, percentages and fractions.

To convert a fraction to a different form, think of it as a division problem that has not yet been divided. Once you divide the numerator (top) by the denominator (bottom) it will be in decimal form. Changing between a decimal and percent is a shift of the decimal point. If you have a decimal number and want to convert it to a percentage, you move the decimal two places to the right and attach a percent sign. Reverse it if you want to go back to the decimal from a percent by moving the decimal two places to the left in the number and removing the percent sign. Converting between fraction and percent just pauses as a decimal number between the two forms.

Examples:

Express $\dfrac{5}{12}$ as a percent.

Express $\dfrac{3}{8}\%$ as a decimal number.

Express 0.33 as a percent.

Express 0.33% as a decimal.

Solutions:

To convert a fraction to a percent, do the division. Here we have 5 divided by 12 so we get $0.41666\overline{6}$ and we can round this to 0.4167. Now we move the decimal two places to the right and attach a percentage sign to get 41.67%

For the next example it is a fraction that is also a percentage. First we will need to divide the fraction so that we can then move the decimal two places to the left and drop the percentage sign. $\frac{3}{8}\% = .375\%$. Now we can move the decimal place to get 0.00375 as the equivalent decimal number.

To convert 0.33 to a percent we shift the decimal two places to the right and add a percentage sign to get 33%.

To convert 0.33% to a decimal number we shift the decimal point two places to the left and drop the percent sign and have 0.0033.

These conversions are important because in most of the problems we will work with in this chapter we will want the numbers to be in decimal form. We will never do calculations with a number in percent form and it is usually easier to convert a fraction to a decimal before doing any calculations. For example, if we want to purchase an item, we can determine the amount we will owe in taxes by converting the tax percent to a decimal and then multiply it by the cost of the item. Or, if the item is on sale, we can determine the discount amount the same way, by multiplying the percent in decimal form by the original price. We would then add the tax amount to the original price, or subtract the discount amount from the original price to determine the cost of the item.

A common equation that uses percentages and decimals together fits the form of A = PB. This might also look like A is what percent of B? In order to work with a problem like this, you will should recognize that "is" represents equals, "what" represents x, and "of" represents multiplication.

Example:

a. 3 is what percent of 10?
b. What is 30% of 10?
c. 3 is 30% of what number?

Solution:

a. $3 = x*10$ or $3=10x$ so $x=3/10 = .3 = 30\%$
b. $x = .3(10) = 3$
c. $3 = .3(x)$ or $3/.3 = x$ or $10 = x$

Another calculation that might use percentages is "percent increase" or "percent decrease". In either case, it is the amount of the increase, or the amount of the decrease, divided by the original amount.

Example:

Suppose tuition increased from $189 per credit hour to $197.50, what percent is the increase?

Solution:

For percent increase or percent decrease, take the amount of the change divided by the original amount. In this case, the amount of increase is 197.50 – 189 = 8.5 so we divide 8.5 by the original cost per credit hour of 189 to get $\frac{8.5}{189} = .04497 \approx .045 = 4.5\%$. That means tuition increased by 4.5%.

Dimensional Analysis

For the rest of this chapter we will focus on converting values within the English measurement system, within the Metric system, and between systems. This process is called **dimensional analysis**.

Within the English system, and between systems, a conversion factor will be used to cancel one set of units and replace it with another. The conversion factor is a ratio that comes from an equality so that it represents the value of one. For example, because 12 inches = 1 ft, $\frac{12\ inches}{1\ foot} = \frac{1\ foot}{12\ inches} = 1$. If we want to convert feet to inches, we will first put our given value over 1, to make it a fraction. Then, choose the conversion factor $\frac{12\ inches}{1\ foot}$ so the foot units cancel once we multiply, and we are left with the equivalent number of inches. Multiplying by the conversion factor did not change the value since it was just multiplying by a form of the number one, but it did change the units when they canceled.

Example:

Convert 29 inches to ft.

Solution:

Write 29 inches as a fraction to get $\frac{29\ inches}{1}$. Use the conversion factor that has inches in the denominator so the inch units will cancel and leave the units of ft.

$$\frac{29\ inches}{1} * \frac{1\ ft}{12\ inches} = \frac{29 * 1ft}{12} = \frac{29}{12}ft = 2\frac{5}{12}ft$$

Below are several equivalent values that can be used for conversion factors.

12 inches = 1 foot

3 feet = 1 yard

5280 feet = 1 mile

60 seconds = 1 minute

Example:

a. Convert 39 inches to yards
b. Convert 27 yards to feet
c. Convert 29,790 feet to miles

Solution:

a. 39 in. = $\dfrac{39\ in.}{1}$

12 inches = 1 foot so $\dfrac{1\ ft.}{12\ in.} = 1$

3 ft. = 1 yd. so $\dfrac{1\ yd.}{3\ ft} = 1$

Now $\dfrac{39\ in}{1} * \dfrac{1\ ft}{12\ in} * \dfrac{1\ yd}{3\ ft} = \dfrac{39}{12*3}yd = \dfrac{39}{36}yd = \dfrac{13}{12}yd$

b. 27 yd = $\dfrac{27\ yd}{1}$

1 yd = 3 ft so $\dfrac{3\ ft}{1yd} = 1$

Now $\dfrac{27\ yd}{1} * \dfrac{3\ ft}{1\ yd} = 27 * 3ft = 81ft$

c. 29,970 ft = $\dfrac{29,970\ ft}{1}$

1 mile = 5280 ft so $\dfrac{1\ mi}{5280\ ft} = 1$

Now $\dfrac{29,970\ ft}{1} * \dfrac{1\ mi}{5280\ ft} = \dfrac{29,790}{5280}mi = \dfrac{993}{176}mi$

Converting units within the Metric system is a very convenient process because the Metric system is a base 10 system. As a result, conversions are based on powers of ten. That means conversions between units are all based on powers of 10, so the process is to shift the decimal place as many units to the right or left that the final unit is away from the initial unit.

Multiply by 10 for each step to the right.

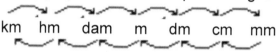

km   hm   dam   m   dm   cm   mm

Divide by 10 for each step to the left.

Metric units represented in numeric form are listed below.

1 kilometer (km) = 1000 meters
1 hectometer (hm) = 100 meters
1 decameter (dam) = 10 meters
1 meter (m) = 1 meter
1 decimeter (dm) = 1/10 meters
1 centimeter (cm) = 1/100 meters
1 millimeter (mm) = 1/1000 meters

Example:

a. Convert 285.7 cm to hm
b. Convert 0.025 mm to m
c. Convert 3358 dm to mm
d. Convert 44g to cg

Solution:

a. Since hm is four units to the left of cm, move the decimal in 285.7 four units left and fill in any spaces with zeros to get:

285.7 cm = 0.02857 hm.

b. Since m is three units to the left of mm, move the decimal in 0.025 three units to the left and fill in as needed with zeros to get 0.025mm = 0.000025m.

c. Since mm is two units to the right of dm, move the decimal in 3358 to the right two places and fill in with zeros to get 3358dm = 335,800mm.

d. Notice this problem is grams rather than meters? Moving by powers of ten still works, just replace the m in the table to grams to see that cg is two units to the right of g so 44g = 4400cg.

Conversions between the English and Metric system work similar to conversions within the English system except the conversion factors come from equalities between the systems such as the approximations below:

1 in   =   2.54 cm
1 ft   =   30.48 cm
1 yd   =   0.9 m
1 mi   =   1.6 km

Sometimes we have to do more than one conversion at a time. For example, converting 14 ft to m we might first convert ft to cm using the list for a conversion factor, but then we will have to move the decimal two places to the left to get from cm to meters.

$$14 \text{ ft} = \frac{14 \, ft}{1}$$

$$1 \text{ ft} = 30.48 \text{ cm so } \frac{30.48 \, cm}{1 \, ft} = 1$$

Now $\frac{14 \, ft}{1} = \frac{30.48 \, cm}{1 \, ft} = 426.72 \, cm = 4.2672 \, m$

Sometimes a problem has two sets of units and we need to use two conversion factors like we did going from inches to yards.

Example:

Express 135 miles per hour in feet per second.

Solution:

$$135 \text{ miles per hour} = \frac{135 \, mi}{1 \, hr}$$

$$5280 \text{ ft} = 1 \text{ mi so } \frac{5280 \, ft}{1 \, mi} = 1$$

$$1 \text{ hr} = 60 \text{ min, so } \frac{60 \, min}{1 \, hr}$$

$$1 \text{ min} = 60 \text{ sec, so } \frac{60 \, sec}{1 \, min}$$

Notice miles in the bottom of the first conversion factor so it will cancel with the miles in the numerator of the problem while hr is in the top of our second conversion factor so it will cancel with the hr in the denominator of our problem.

Now

$$\frac{135\ mi}{1\ hr} * \frac{5280\ ft}{1\ mi} * \frac{1\ hr}{60\ min} * \frac{1\ min}{60\ sec} = \frac{135 * 5280ft}{3600\ sec} = 198ft/sec$$

Notice, to get a table of equalities for values that represent area, or squared values, we simply square both sides of our prior tables so that if:

    3 ft = 1 yd, then   1 in = 2.54 cm, then
    (3ft)(3ft) = 1 yd²   1 in²= (2.54cm)(2.54cm)
    9 ft²=1 yd²   1 in²=6.5 cm²

Once you have conversion factors from the equalities for squared values, the same process is used for conversions. A table of conversion equalities can also be used for cubic or volume quantities.

1 in² = 6.5 cm²
1 ft² = 0.09 m²
1 yd² = 0.8 m²
1 mi² = 2.6 km²
1 acre = 0.4 hectare (ha)
2 pints (pt) = 1 quart (qt)
4 quarts = 1 gallon (gal)
1 gallon = 128 ounces (oz)
1 cup (c) = 8 ounces
1 yd³ ≈ 200 gallons
1 ft³ ≈ 7.48 gallons
231 in³ ≈ 1 gallon
1 cm³ = 1 mL
1 dm³³ = 1000 cm = 1L
1 m = 1 kL
16 oz = 1 lb

2000 lb = 1 T
1 oz = 28 g
1 lb = 0.45 kg
1 T = 0.9t

Example:

a. Convert 2.8 m² to yd²
b. Convert 7 in² to cm²
c. Convert 35,000 ft³ to gal
d. Convert 3300 gal to yd³

Solution:

a. $2.8 \ m^2 = \dfrac{2.8 \ m^2}{1}$

$1 \ yd^2 = 0.8 \ m^2$ so $\dfrac{1 \ yd^2}{0.8 \ m^2} = 1$

Now $\dfrac{2.8 \ m^2}{1} * \dfrac{1 \ yd^2}{0.8 \ m^2} = \dfrac{28}{8}yd^2 = \dfrac{7}{2}yd^2 = 3.5yd^2$

b. $7 \ in^2 = \dfrac{7 \ in^2}{1}$

$1 \ in^2 = 6.5 \ cm^2$ so $\dfrac{6.5 \ cm^2}{1 \ in^2} = 1$

Now $\dfrac{7 \ in^2}{1} * \dfrac{6.5 \ cm^2}{1 \ in^2} = 7 * 6.5 \ cm^2 = 45.5 \ cm^2$

c. $35,000 \ ft^3 = \dfrac{35,000 \ ft^3}{1}$

$1 \ ft^3 = 7.48$ gallons so $\dfrac{7.48 \ gal}{1ft^3} = 1$

$$\text{Now} \quad \frac{35,000\ ft^3}{1} * \frac{7.48\ gal}{1 ft^3} = 35,000 * 7.48\ gal = 261,800\ gal$$

d. 3300 gallons = $\dfrac{3300\ gal}{1}$

1 yd³ = 200 gal so $\dfrac{1\ yd^3}{200\ gal} = 1$

$$\text{Now} \quad \frac{3300\ gal}{1} * \frac{1\ yd^3}{200\ gal} = \frac{33}{2}yd^3$$

This conversion process works whenever we have an equality that can be used for the conversion. That means 490 lb. can be converted to kg if we know the equality 1 lb. = 0.45 kg

$$\text{We have:} \quad \frac{490\ lb}{1} * \frac{0.45\ kg}{1\ lb} = 490 * 0.45\ kg = 220.5\ kg$$

Temperature
Temperatures can also be converted but an equation is used to relate Celsius and Fahrenheit.

C = (F − 32) * $\dfrac{5}{9}$

F = C* $\dfrac{9}{5}$ + 32

Example:

a. Convert -1°C to Fahrenheit
b. Convert 49° F to Celsius

Solution:

a. $F = C * \dfrac{5}{9} + 32$

$F = -1 * \dfrac{9}{5} + 32$

$F = -\dfrac{9}{5} + 32$

$F = 30.2°$

b. $C = (F - 32) * \dfrac{5}{9}$

$C = (49-32) \dfrac{5}{9}$

$C = 17(\dfrac{5}{9})$

$C = \dfrac{85}{9} = 9.4°$

Chapter Review

Dimensional Analysis – the process of converting the units of a measurement within the English or metric system, or between these measurement systems.

Metric System – the system of measurement used in the majority of the world that includes meters for length, grams for weight and centigrade for temperature.

English System – the system of measurement originating in England used in the United States and other countries that includes feet and inches for length, pounds for weight and Fahrenheit for temperature.

Explorations

In the News: Measurement Mania

Find a newspaper article that converts measurements at some point in the article. Show the calculations that are needed to convert the measurements and indicate whether the published results appear to have been calculated correctly.

Writing Across the curriculum

1. Write a report about Fahrenheit, Celsius, and Calvin temperature scales such as when each is used and why we have different temperature scales. Be sure to include how these different systems affect the values of community and respect.

2. Compare and contrast Metric and English measurement systems. Also include where each is used and why we have different measurement systems. What reasons might be given for continuing to use the English system? How does the use of the English or Metric system affect community?

3. Compare and contrast conversions in linear measurement, area measurement, and volume measurement.

4. Write a paper describing why every measurement has two parts.

Exercise 1

The Google search engine (www.google.com) will convert from one unit to another. For example, to convert miles to kilometers you would type 10 miles to kilometers in the search box. Visit Google and convert the following: 10 light years to parsecs, 1 dollar to yen, 8.5 teaspoons to ounces, 6 hectares to acres.

Discussion

1. Find something that has the measurement reported in Metric measurement, express why it is reported that way, and express the measurement in two equivalent ways using the English measurement system.

2. Find a reported measurement and express the measurement three different ways. Which way is most effective and why?

3. How might a measurement be interpreted in different ways? What affect can this have on community?

4. Why might the ability to estimate be important in understanding measurements? What are some problems you might encounter if you are unable to estimate measurements?

Chapter 3: Geometry Basics

There are shapes all around us. As you read these next two chapters, look around you and notice the shapes, figures, and dimensions such as perimeter, area, and volume, on everything you see. These figures are made up of lines, line segments, and angles. A **line** goes on infinitely in both directions, while a **line segment** has a finite end, so you can see a line segment while a line does not stop with your line of sight. When you connect line segments, they form an **angle**, or distance rotated between the two line segments.

Angles

If one line segment rotates all the way around until it is back where it started, it has rotated through 360° degrees. That means two line segments connected end to end to form a single, longer line segment, is connected with a 180° angle, sometimes called a **straight angle**. A square has four corners that are 90° each, these are called **right angles**. An angle that is smaller than a right angle is called an **acute angle**, so acute angles fall between 0° and 90°, 0°< acute angle < 90°. If an angle is larger than a right angle, it is called obtuse. An **obtuse angle** is between 90° and 180°. 90° < obtuse angle < 180°.

Example:

Determine how many degrees the hour hand of a clock moves when the time changes from:

1. 10 o'clock to midnight
2. 2 o'clock to 7 o'clock

Solution:

a. Since there are 12 hours on the clock, each hour represents $\frac{360}{12} = 30°$. There are 2 hours between 10:00 and midnight so 2*30 = 60°

b. There are 5 hours between 2:00 and 7:00 for a total of 5*30 = 150°.

Example:

Use the protractor to find the measurement of ∠DBC then identify whether the angle is acute, right, straight, or obtuse.

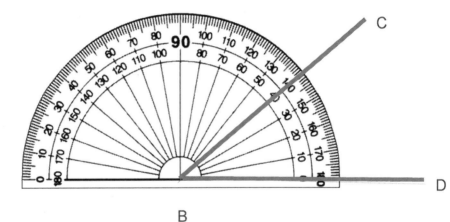

B

Solution:

The angle rotates through 40° and is acute because 0 < 40° < 90°

Two angles that add up to a straight angle, or 180° are called **supplementary angles**. The angles are easy to identify when they are together and form a straight angle, however, they do not have to be connected to be considered supplementary angles, they just have to add up to 180°. If the sum of two angles is 90° then they are called **complementary angles**.

Example:

Find the measure of the missing angle.

Solution:

Because these two angles form a straight angle, or add up to 180°, they are supplementary angles. Since 180° - 44° = 136°, 136°is the measure of the missing angle.

Example:

Find the measure of the complement and supplement of 27.3°.

Solution:

The complement will add to 27.3° to make 90° so 90 – 27.3 = 62.7°

The supplement will add to 27.3° to make 180° so 180 – 27.3 = 152.7°

Angle Definitions:

When lines cross, they form pairs of angles that are opposite each other. These pairs of angles are called **vertical angles** and are equal in measure. When two parallel lines are cut by another line, that line is called a **transversal**. Several interesting pairs of equal angles are formed. There are four sets of vertical angles that are formed, $\angle 1 = \angle 4$, $\angle 2 = \angle 3$, $\angle 5 = \angle 8$, $\angle 6 = \angle 7$. In addition, there are the **alternate,** (or opposite) **interior angles** that are equal in measure $\angle 4 = \angle 5$, $\angle 6 = \angle 3$, the **alternate exterior angles** $\angle 7 = \angle 2$, $\angle 8 = \angle 1$, and the **corresponding angles**, or angles in the same position on the two parallel lines are equal $\angle 2 = \angle 6$, $\angle 4 = \angle 8$, $\angle 3 = \angle 7$, $\angle 1 = \angle 5$.

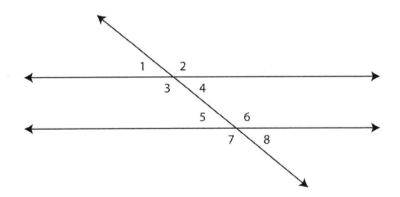

Example:

Find the measure of the missing angles if angle 1 is 39°

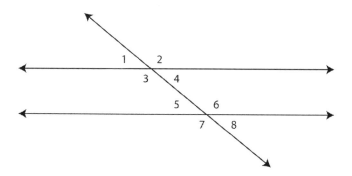

Solution:

$\angle 1 = 39°$ is given.

$\angle 2$ and $\angle 1$ add to 180° so $\angle 2 = 180 - 39 = 141°$.

$\angle 3 = \angle 2$ because they are vertical angles so $\angle 3 = 141°$.

$\angle 4 = \angle 1$ because they are vertical angles so $\angle 4 = 39°$.

$\angle 5 = \angle 4$ because they are alternate interior angles and $\angle 5 = \angle 1$ because they are corresponding angles, so either way, $\angle 5 = 39°$.

$\angle 6 = \angle 2$ because they are corresponding angles, $\angle 6 = \angle 3$ because they are alternate interior angles, and $\angle 6 + \angle 5 = 180°$. All of these situations result in $\angle 6 = 141°$.

∠7 = ∠6 = 141° because they are vertical angles.

∠8 = ∠5 = 39° because they are vertical angles.

Triangles

Angles are a great way to transition into triangles because the three angles in a triangle always add up to 180°. There are several types of triangles. If one of the angels is 90° then it is a called **right triangle**. A small box drawn in an angle is the symbol to represent a right angle since the naked eye cannot distinguish if an angle is exactly 90° or one or two degrees different. If the three angles are all the same measure, or 60°, then the three sides are also the same measure and the triangle is called an **equilateral triangle**. If two of the angels have the same measure, but the third is different, then two if the sides also have the same measure and the triangle is called an **isosceles triangle**. If all angles, and therefore sides, are a different measure, it is called a **scalene triangle**.

Although many triangles might fit each of these categories, they are only considered to be **similar triangles** if the measure of all three angles from one triangle is the same as the measure of all three angles in another triangle. When two triangles are similar, it means the three ratios of the lengths of the corresponding sides in the two triangles are all equal. It also means the ratio of any two sides in one triangle will be equal to the ratio of the corresponding sides in the similar triangle.

Example:

Find the measurement of the missing angles.

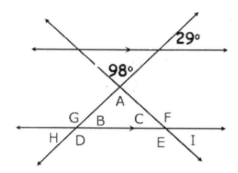

Solution:

Since the angle opposite ∠A is 98°, ∠A = 98°.

The corresponding angle to ∠B is 29° so ∠B = 29°.

Since ∠A + ∠B + ∠C = 180° as a triangle, ∠C = 180 – 98 – 29 = 53°.

∠D + ∠B = 180° so ∠D = 180 – 29 = 151°.

∠E + ∠C = 180° so ∠E = 180 – 53 = 127°.

∠F is vertical to ∠ E so ∠F = 127°.

∠G is vertical to ∠D so ∠G = 151°.

∠H is vertical to ∠B so ∠H = 29°.

∠I is vertical to ∠C so ∠I = 53°.

Example:

Find the measure of the missing angle.

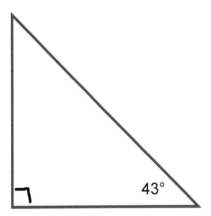

Solution:

The three angles of a triangle add up to 180° so the missing angle is 180 – 90 – 43 = 47°

Pythagorean Theorem
In any right triangle, the side opposite the right angle is called the hypotenuse and the other two sides are called the legs. The Pythagorean theorem states that the square of the length of the hypotenuse will always equal the sum of the squares of the two legs.

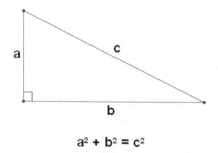

$$a^2 + b^2 = c^2$$

Example:

Use the Pythagorean theorem to find the missing length in the right triangle.

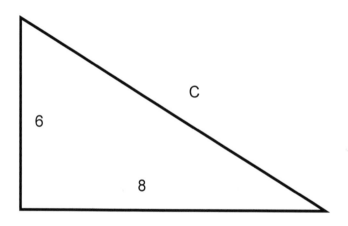

Solution:

$c^2 = 6^2 + 8^2 = 36 + 64 = 100$

$c = \sqrt{100} = 10$

Polygon

We have seen that a three-sided figure is called a triangle. A triangle is also a polynomial because any closed figure with straight edges is a **polygon**. To be **closed** simply means the straight lines meet at **vertices**, or points of intersection, so you can trace the entire figure without lifting your pencil. A four-sided polynomial is also called a **quadrilateral**. A five-sided polynomial is a **pentagon**. A six-sided polynomial is called a **hexagon** while a seven-sided polynomial is a **heptagon**. We will finish with the **Octagon**, or eight-sided figure. Anything higher can be found in a quick online search.

If the lengths of all sides are the same, and therefore, the measures of all angles are equal, the figure is called a **regular** polygon. The sum of the measure of all angles in a regular polygon is 180°(n-2) where n is the number of sides in the figure. Just divide the result by the number of sides to find the measure of each individual angle.

There are several types of quadrilaterals. Another name for a regular quadrilateral is the **square** because all four sides are equal and all four angles are 90°. If the angles remain equal, but the sides are no longer all equal in length, it is a **rectangle**. If the sides remain equal, but the angles are no longer 90° then it is a **rhombus**. If two sets of sides are parallel, then it is a **parallelogram**. That means a square, a rhombus, and a rectangle can all also be considered to be a parallelogram as well as a quadrilateral and a polygon. Finally, if the quadrilateral has one set of parallel sides, the figure is a **trapezoid**.

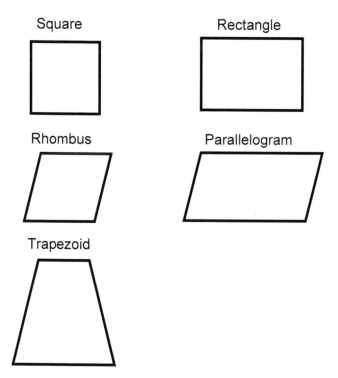

Square

Rectangle

Rhombus

Parallelogram

Trapezoid

Regardless of the type of Polynomial, the way to find the perimeter is always the same. Since the **Perimeter** is the length around the outside of the polygon, you add up the length of all the sides to get the perimeter.

Example:

Find the perimeter of the following figure.

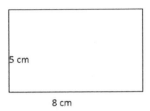

5 cm

8 cm

Solution:

Since this is a rectangle, opposite sides are equal length, so the perimeter is:

5cm + 5cm + 8cm + 8cm = 26cm

Example:

Find the perimeter of the polygon below:

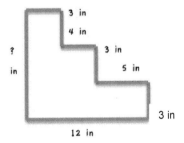

3 in

4 in

3 in

? in

5 in

3 in

12 in

Solution:

Add the length of all the sides. Notice the length of the left side is missing? If we add the right side up, that will equal the length of the missing side. That means the left side is equal to 3in + 3in + 3in = 9in. The top is also missing, but if we take the bottom measurement of 12 inches and subtract the 5 inches and 4 inches above, the remaining top will be 12in − 5in − 4in = 3in

Perimeter = 3in + 4in + 3in + 5in + 3in + 12in + 9in + 3in = 42in

Example:

A landscape artist is designing a xeriscape garden for a house in Florida. The space is two times as long as it is wide. The perimeter is 120 yards. He plans to plan flowers around the outer edge, spaced 2 feet apart along both sides of the length. How many flowers will he need?

Solution:

First let x = width. Then length = 2x.

Perimeter = 120 = 2x + 2x + x + x = 6x so x = 120/6 = 20 yards. Since x is width, and length = 2x, that means length = 2* 20 yards = 40 yards.

There are two lengths, so flowers will fill 2 * 40 yards = 80 yards.

The flower distance is given in feet so we multiply 80 yards by 3 feet to get 80 yards * $\dfrac{3\,feet}{1\,yard}$ = 240 feet.

The flowers will go 2 feet apart so he will need 240/2 = 120 flowers to fill the space.

Chapter Review

Lines & Angles

Line	Goes infinitely in both directions
Line Segment	Has a finite beginning and end
Angle	Two line segments intersect to form an angle
Straight Angle	A straight line, angle of 180 degrees
Right Angle	Square, or 90 degree angles
Acute Angle	An angle less than 90 degrees
Obtuse Angle	An angle greater than 90 degrees, less than 180 degrees
Supplementary Angles	Two angles that add up to 180 degrees
Complementary Angles	Two angles that add up to 90 degrees
Vertical Angles	Two equal angles created by intersecting lines
Interior Angles	Equal Interior angles created by a line intersecting parallel lines

Exterior Angles	Equal exterior angles created by a line intersecting parallel lines
Corresponding Angles	Two angles in the same position when intersecting parallel lines

Triangles

Triangle	Polygon with three sides
Right Triangle	Triangle with one 90° angle
Equilateral Triangle	Triangle with 3 equal sides and 60° angles
Isosceles Triangle	Triangle with two equal sides
Scalene Triangle	Triangle with 3 different length sides
Similar Triangles	Two triangles with three angles that are equal
Pythagorean Theorem	The square of the length of the hypotenuse will equal the sum of the squares of the two legs.

Polygons

Polygon	A closed two-dimensional shape with straight edges. Closed indicating the edges meet at vertices.
Quadrilateral	A four sided polygon

Pentagon	A five sided polygon
Hexagon	A six sided polygon
Heptagon	A seven sided polygon
Octagon	An eight sided polygon
Regular Polygon	A polygon where all angles are equal
Square	A regular quadrilateral with four equal sides
Rectangle	A regular quadrilateral
Rhombus	A quadrilateral with four equal sides, but the angles are not 90°
Parallelogram	A quadrilateral where two sets of sides are parallel
Trapezoid	A quadrilateral with one set of parallel sides
Perimeter	Length around the outside of the polygon

Explorations

In The News: Geometry Gems

Choose a particular geometric figure and find at least 3 newspaper articles that use, describe, or illustrate that figure in some way. Report with as much information as you can about the figure as well as how it was used. How does the figure affect Community?

Writing Across the Curriculum

1. Research an artist who utilizes geometry. Write a report about the artist and discuss how geometry is used in his/her work.

2. What is your favorite shape and why? Please describe the features of the shape as specifically as possible.

3. Why do you think most signs are modeled after mathematical shapes rather than artistic swirls and other designs? How does this affect the community of drivers?

4. How is the study of Geometry similar to the study of Algebra? How is it different?

5. How is the study of Geometry similar to Measurement? How is it different?

6. Reference an artistic work that clearly utilizes geometry. Describe the geometry that is used.

7. How do architects use geometry? How does that affect community?

Critical Thinking

1. Create a geometric figure, illustrate all relevant measurements, and find the perimeter and area.

2. Find a geometric figure, take a picture, label all relevant measurements, and find the perimeter and area.

3. Find two geometric figures that were not mentioned in this chapter. Express all relevant facts about the figures.

4. How might our community be different without Geometry?

Chapter 4: Geometry

Geometry is an important area of mathematics used in design, architecture, fashion and many other industries. The key areas of geometry covered in this section include area, circles, volume, and surface area.

Area

There are different types of area, but they are calculated the same. For example, there is the area of a two dimensional figure, and the area of the surfaces of a three-dimensional figure. The surface areas of the three dimensional figure are made up of two-dimensional figures that you then add together. The area of many figures can be configured with an understanding of key features contained in the figures. We will start with the area of a basic figure that most people are familiar with. The area of a **rectangle** is the length times the width, or

$$A_{rectangle} = l*w$$

If we understand this formula, we can follow where the formula for the area of a square comes from. The area of a **square** would be either L * L or W * W since the length is equal to the width. Or, we can just call the length of the sides S, since they are all the same length and the area of the rectangle then becomes the equivalent area for the square of side times side or

$A_{square} = s * s = s^2$

A triangle can be constructed by cutting a rectangle in half. As a result, the area of a triangle would be half of the length times the width. However, there are three sides to a triangle so we need to have a way to determine which side is the length and which is the width. If you start at any of the vertices in a triangle, and drop a line straight down so that it is perpendicular to the opposite side, those two values serve as the length and width for the purpose of the area of a **triangle** and length and width become height and base to get the formula

$$A_{triangle} = \frac{1}{2}bh$$

If you cut a diagonal across a **rhombus**, you have two triangles. Both of the triangles will have the same length base from the definition of a rhombus. As a result, the area of each triangle will be the same. Combining those two areas gives twice the area of a triangle or

$$A_{rhombus} = bh$$

The formula that area = base X height holds true for most quadrilaterals. In a square, we already saw that the area = s^2. Since base and height are both equal to s, we could use the formula A = bh instead. In a rectangle the base and height are the same as the length and width, so A = bh applies again. The same will hold true for all **parallelograms** and we have

$$A_{parallelogram} = bh$$

The **trapezoid** is one quadrilateral that does not fit this pattern. That is because the base has different values depending which base is used. As a result, if you average the two bases together, then the same formula holds true and

$$A_{trapezoid} = \frac{b1 + b2}{2} * h$$

Example:

Find the area of the following figures:

5 cm

8 cm

a.

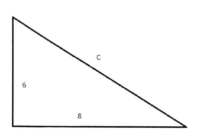

b.

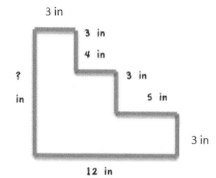

c.

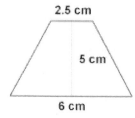

d.

Solution:

a. $A_{rectangle} = l*w = 5cm*8cm = 40cm^2$

b. $A_{triangle} = \frac{1}{2}bh = \frac{1}{2}*8*6$ square units = 24 square units

c. This one breaks into 3 quadrilaterals. If I break a 3in X 3in square off the top, then the next rectangle down will be 3in + 4in wide and 3in tall or 7in X 3 in and the bottom rectangle is 3in + 4in + 5in wide and 3 in tall or 12in X 3in for a total of:

A = 3in X 3in + 7in X 3in + 12in X 3in
A = 9 in² + 21 in² + 36 in²
A = 66 in²

d. $A_{trapezoid} = \dfrac{b1 + b2}{2} * h = \dfrac{2.5cm + 6cm}{2} * 5cm = \dfrac{8.5\,cm}{2} * 5cm$ = 4.25cm*5cm = 21.25cm²

Circles

We have looked at perimeter and area for polygons, but circles are not polygons. A **circle** is a figure where every point on the figure is an equal distance from one point, called the center. The distance that each point on the circle is away from the center is called the **radius**. The **diameter** is the distance across the circle, crossing through the center. Since the line segment across the circle, which passes through the center, is made up of two radii, we have:

D = 2r

The term perimeter is not used with circles, the distance around is instead called the circumference and is given by

$C = 2\pi r$

The area of a circle cannot use the formulas from the quadrilaterals since there is no base or height in a circle. Instead we have

$A_{circle} = \pi r^2$

Example:

Find the area, circumference, and diameter of the circle.

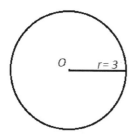

Solution:

$D = 2r = 2*3 = 6$
$C = 2\pi r = 2\pi * 3 = 6\pi$
$A_{circle} = \pi r^2 = \pi * 3^2 = 9\pi$

Volume

Volume is the cubic quantity a figure holds. It is cubic because you are multiplying three dimensions to get the result, where square area is two-dimensional. A general idea for volume is that you take the area of the base and multiply it by the height. For example, you have probably heard of the volume of a rectangular solid as length times width times height.

The area of the base is length times width so you are now multiplying by the height to get volume.

$V_{\text{rectangular solid}} = l*w*h$

For a cube, this formula can be simplified because the length = width = height to get

$V_{\text{cube}} = s^3$

A cylinder would also work this way. The base of a circular cylinder is a circle, so you have the area of the circle times the height or

$V_{\text{circular cylinder}} = \pi r^2 h$

Some shapes do not have sides that go up at a direction perpendicular to the base. Some rise to a point at the top, such as a cone or a rectangular pyramid. In cases like that, the volume is 1/3 of the value it would be if the sides were perpendicular. In other words, the volume of a cone is 1/3 the volume of a right circular cylinder. The volume of a rectangular pyramid is 1/3 the volume of a rectangular solid.

$V_{\text{cone}} = 1/3 \; \pi r^2 h$

$V_{\text{rectangular pyramid}} = 1/3 \; lwh$

Just as circles fit a different form than the other two-dimensional figures, spheres stand alone among the three dimensional shapes. The volume of a sphere is given as

$V_{\text{sphere}} = 4/3 \; \pi r^3$

Example:

Find the volume of the following figures.

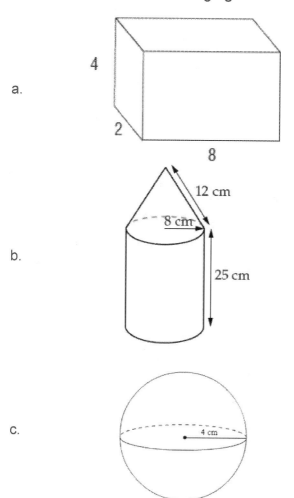

a.

4

2

8

12 cm

8 cm

b.

25 cm

c.

4 cm

Solution:

a. This is a rectangular solid, so we have
$V_{rectangular\ solid} = l*w*h = 2*4*8 = 64$ square units

b. There are two shapes here. We will need to find the volume of the cone and the volume of the cylinder.

$V_{cone} = 1/3\ \pi r^2 h$

The radius is 8cm, but we are not given the height. Using the Pythagorean theorem:

8cm^2 + h^2 = 12cm^2
64cm^2 + h^2 = 144cm^2
h^2= 80cm^2
h = $\sqrt{80}$cm = $4\sqrt{5}$cm.

That gives the volume of the cone as:

V_{cone} = $1/3\ \pi 8^2 4\sqrt{5}$

V_{cone} = $\dfrac{64 * 4\pi\sqrt{5}}{3}$

V_{cone} = $\dfrac{256\pi\sqrt{5}}{3}$ cm^3

$V_{circular\ cylinder}$ = $\pi r^2 h$
$V_{circular\ cylinder}$ = $\pi 8^2(25)$
$V_{circular\ cylinder}$ = 1600πcm^3

So the volume of the whole figure is:
V = V_{cone} + $V_{circular\ cylinder}$
V = $(1600\pi + \dfrac{256\pi\sqrt{5}}{3})$ cm^3

c. This is the volume of a sphere with radius 4cm
V_{sphere} = $4/3\ \pi r^3$ = $4/3\ \pi 4^3$ = $\dfrac{256\pi}{3}$

Surface Area

Finding surface area is the same as finding area of all the faces of a solid. You determine what the different surfaces are on the figure, find the area of each, and add them together to get the surface area. In a cube, each surface is a square with side length s. There are 6 sides on a cube so you have the area of the six squares or the surface area of a cube is equal to $6s^2$.

$SA_{cube} = 6s^2$

The surface area of a rectangular solid is similar, except there are 3 different shaped surfaces. You have the left and right side, the top and bottom, and the front and back surfaces. These surfaces have areas of lw, wh, and lh where l is the length, w is the width, and h is the height. There are two surfaces with each of these areas for a total surface area of 2lw + 2lh + 2wh.

$SA_{rectangular\ solid} = 2lw + 2lh + 2wh$

A right circular cylinder can be broken into two different shapes. The ends are circles and if you cut it down the side to open up, the middle is a rectangle that has length as the height of the cylinder, and the width is the distance around the circle, or $2\pi rh$. That means the surface area is equal to the area of the two circles and the rectangle, or $2\pi r^2 + 2\pi rh$.

$SA_{right\ circular\ cylinder} = 2\pi r^2 + 2\pi rh$

Example:

Find the surface area of the following figures.

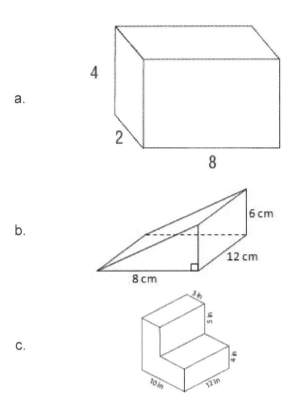

a.

b.

c.

Solution

a. This is a rectangular solid where w = 2, l = 8, and height = 4

$SA_{rectangular\ solid}$ = 2lw + 2lh + 2wh

SA = 2(8)(2) + 2(8)(4) + 2(2)(4)

SA = 32 + 64 + 16

SA = 112 square units

b. This figure has 2 ends that are identical triangles, and five different surfaces.

The area of each triangle is ½ * 6cm * 8cm = 24cm².

The area of the base rectangle is 8cm*12cm = 96cm^2.

The area of the front rectangle is 12cm*6cm = 72cm^2.

To find the area of the slanted rectangle we use the Pythagorean theorem with $(8cm)^2 + (6cm)^2 = c^2$ so 64cm^2 + 36cm^2 = c^2 combining like terms we have 100cm^2 = c^2, taking the square root of each side results in 10cm = c. The sides of the slanted rectangle are 10cm and 12cm so the area is 10cm * 12cm = 120 cm^2.

For the surface area of the figure we add the two triangles and three sides to get:

SA = 2(24cm^2) + 96cm^2 + 72cm^2 + 120cm^2

SA = 336cm^2.

c. This figure has 8 sides; can you find them all? To find the surface area of all the sides we need to see that the depth of the first step is equal to the total depth of 10in minus the 3in depth of the top step, for a total of 7in. We will also need the total height of the back and can get that by adding the 4in and 5in steps for a total height of 9in. Now we are ready to find the 8 sides.

The front and back side are identical and will take us multiple steps so let's start there. If we first calculate the complete bottom rectangular portion of the L-shape, the area is 10in*4in = 40in^2.

That leaves the smaller top part of the L-shape with an area of 3in*5in = 15in². Together the area of the whole L-shape is 40in² + 15in² = 55in².

The area of the bottom of the figure is 12in*10in = 120in².

The backside of the steps is 10in*9in = 90in².

Starting at the bottom, and working up the stairs, the rectangles have area of 12in*4in = 48in², 12in*7in = 84in², 12in*5in = 60in², and finally 12in*3in = 36in².

Now, to add up the area of the eight sides: 55in² + 55in² + 120in² + 90in² + 48in² + 84in² + 60in² + 36in² = 548in².

Example:

Jim is going to resurface his pool. The cost is $3.83 per square foot. The pool is 10 feet wide, 18 feet long, 3 feet deep at the shallow end, and slopes to 6 feet deep in the deep end. How much will it cost him to resurface his pool?

Solution:

The total cost will be the surface area times the cost per foot to resurface the pool. The top 3 feet of the pool is made up of four rectangular shaped walls. In addition, there will be 2 triangular walls for the bottom half of the sides, where the bottom drops from 3 feet to 6 feet. Finally, the bottom half of the wall on the deep end will be another rectangle that is 3 feet high.

The area of the four upper rectangular regions is made up of two regions that are 3ft*10ft = 30ft² and two regions that are 3ft*18ft = 54ft^2. The bottom rectangle on the deep end is 3ft*10ft = 30ft^2. Finally, the two lower triangular regions are each ½*3ft*18ft = 27ft^2. Adding these seven pieces together the total surface area is 2*30ft^2 + 2*54ft^2 + 30ft^2 + 2*27ft^2 = 252ft^2

Since resurfacing costs $3.83 per square foot, the total cost should be approximately $3.83 * 252 = $965.16.

Chapter Review

Area Formulas

$A_{rectangle} = l*w$

$A_{square} = s * s = s^2$

$A_{triangle} = \dfrac{1}{2}bh$

$A_{parallelogram} = bh$

$A_{trapezoid} = \dfrac{b1 + b2}{2} * h$

Volume Formulas

$V_{rectangular\ solid} = l*w*h$

$V_{cube} = s^3$

$V_{circular\ cylinder} = \pi r^2 h$

$V_{cone} = 1/3\ \pi r^2 h$

$V_{rectangular\ pyramid} = 1/3\ lwh$

$V_{sphere} = 4/3\ \pi r^3$

Circle Equations

$D = 2r$

$C = 2\pi r$

$A_{circle} = \pi r^2$

Surface Area Formulas

$SA_{cube} = 6s^2$

$SA_{rectangular\ solid} = 2lw + 2lh + 2wh$

$SA_{right\ circular\ cylinder} = 2\pi r^2 + 2\pi rh$

Explorations

In The News: Geometry Gems

Find an article that utilizes values for area, volume, or surface area. Check the calculations and show how the total area or volume was computed. (You might have to find an article that references a common item or has the dimensions available).

Exercise 1

1. Find a real example of each of the following, around your home, work, or campus, and describe the example in detail or print a picture:
   - Isosceles triangle
   - Equilateral triangle
   - Scalene triangle
   - Regular polygon
   - Pentagon
   - Hexagon
   - Heptagon
   - Octagon
   - Parallelogram
   - Rhombus
   - Rectangle
   - Square
   - Trapezoid
   - Point
   - Line segment
   - Right angle
   - Acute angle
   - Obtuse angle

- Straight angle
- Vertical angles
- Complimentary angles
- Supplementary angles
- Rectangular solid
- Cube
- Pyramid
- Right circular cylinder
- Cone
- Sphere
- Similar Triangles

2. Find the perimeter of 3 of the above figures.

3. Find the total degree of all angles and individual angle measurements of your regular polygon.

4. Find the area of 3 of your figures.

5. Find the volume of 3 of the above figures.

6. Find the surface area of 2 of the above figures.

7. How might geometry be used to build a feeling of community?

8. How can geometry be used to promote responsible stewardship?

Here is an example of what you should be doing in this exercise:

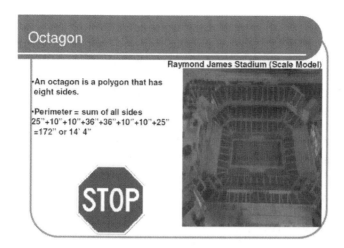

Octagon

Raymond James Stadium (Scale Model)

•An octagon is a polygon that has eight sides.

•Perimeter = sum of all sides
25"+10"+10"+36"+36"+10"+10"+25"
=172" or 14' 4"

STOP

Writing Across the Curriculum

1. Write a report illustrating how geometry is used in industry. Be sure to include how the values of community, responsible stewardship, and excellence affect the industry you researched.

2. Compare and contrast how a designer and architect might use geometry. How does Geometry effect the community from the designer and architect perspective?

3. Bigger is often cheaper per item or amount. For example, buying food in bulk. We see this with the growth in the number of warehouse clubs that sell items in larger quantities for a lower per item cost. What are some situations when bigger is not better, even when it is cheaper.

Exercise 2

Create a three dimensional figure. Label all relevant dimensions. Calculate the area of each face, the number of vertices, the perimeter of each face, the total surface area, and the volume. Express your results in both the English and Metric system.

Critical Thinking:

1. What is one shape you think everyone sees every day? Where are some of the places you think they see it? Who is everyone in your response?

2. What is one shape you think very few people know about, and why?

3. What is one shape your see regularly but that you never noticed until this class?

Chapter 5: Counting and Probability

You probably already thought you knew how to count. Counting falls within the mathematical field of Combinatorics so we will expand on traditional counting in order to solve counting problems. The methods of counting covered in this chapter are permutations and combinations.

In order to be able to use the formulas for permutations and combinations, let's first look at factorial notation. **Factorial notation** is given by an exclamation mark after a constant. It means to multiply that number by every natural number less than it. For example, $3! = 3*2*1 = 6$ and $7! = 7*6*5*4*3*2*1 = 5,040$. In general the formula looks like

$$n! = n*(n-1)*(n-2)*\ldots (3)*(2)*(1)$$

Factorials can be simplified in fractions because the numbers are multiplied together. As a result, if you have the same factorial in the numerator as the denominator, that factorial can be canceled. It is the same as canceling each of the factors within the factorial. For example, if you have $\frac{14!}{(12)!}$ you can break the numerator down into $14*13*12!$ since that is still the product of all the natural numbers below and including 14. Now your fraction becomes: $\frac{14*13*12!}{(12)!}$ and you can cancel the 12! from the numerator and denominator because it is the same as having the common factors of 12, 11, 10, 9, 8, 7, 6, 5, 4, 3, 2 and 1. However, you can cancel all 12 factors in one step by canceling the factorial to be left with 14*13 = 182. This is especially helpful when the factorials are too large for your calculator to handle.

Fundamental Counting Principle

The Fundamental Counting Principle states that if you have m items of one type and n items of another type, there are m*n ways you can combine those items. We can use this principle to determine the number of arrangements that can be made in many different situations.

Example:

How many unique phone numbers can be created for each area code?

Solution:

Each phone number has seven digits after the area code. Each digit can be chosen from the numbers 0 through 9 so there are 10 choices for each digit after the first one. There are only 8 choices for the first digit because you cannot dial 1 or 0 first in a phone number. That means there are 8*10*10*10*10*10*10 = $8*10^6$ = 8,000,000 different phone numbers possible for each area code.

Permutations

Permutations are the number of ways a certain number of objects can be arranged if the order of the arrangement is important, so that each different arrangement gives a different result. In other words, if the order makes a new arrangement, you are working with permutations. Examples would include electing officers in a club. If there is a president, secretary, and treasurer, you might be choosing three officers from a club with 20 members. The order is defined by an office, so that if you choose three students but change the order of those students, you have a new arrangement because the students would be filling a different office. The notation for choosing three officers from a club of 20 members looks like $_{20}P_3$. Meaning the number of permutations of 20 members taken three at a time.

Permutations can be calculated using commands on a graphing calculator, on Excel, or by hand. The formula for the number of permutations of n items taken k at a time is given by:

$$_nP_k = \frac{n!}{(n-k)!}$$

Example:

The 5th grade class at Madeira Fundamental Elementary School has 24 students and must choose a representative and an alternate student government representative. How many different ways can they choose the representative and the alternate?

Solution:

Since there are 24 students two at a time and the order changes whether the student is the representative or alternate, we would calculate the number of permutations of 24 items taken two at a time to have:

$$_{24}P_2 = \frac{24!}{(24-2)!} = \frac{24!}{(22)!} = \frac{24*23*22!}{(22)!} = 24*23 = 552$$ so there are 552 different ways a representative and an alternate can be chosen from the 24 students in the class.

Example:

How many different teams can you put together in volleyball if you have 12 hitters and five play at a time with your setter?

Solution:

Since each arrangement puts different players next to each other and in different positions on the court, each arrangement of five players is different each time you move a player. As a result, this is the same as the number of permutations of 12 players taken five at a time or

$$_{12}P_5 = \frac{12!}{(12-5)!} = \frac{12!}{(7)!} = \frac{12*11*10*9*8*7!}{(7)!} =$$
$12*11*10*9*8 = 95,040$ different player arrangements on the court. Too bad permutations don't tell you what arrangement will win.

These calculations can be done on many graphing calculators, or on Excel using the following command: =PERMUT(12,5) and then select "enter"

Combinations

Next is combinations and then how to determine which to use. Notice with permutations, each arrangement was different if the order of the objects was changed. With **Combinations**, the order of the objects does not matter, only the objects themselves. Combinations might include the number of ways to choose five stocks for your portfolio. The order in which you choose them will not matter, only the fact that you chose each particular group of five stocks. Or, if your group wanted to choose three representatives for a committee. The order of the people chosen will not change the committee of individuals.

Combinations can also be calculated using commands on a graphing calculator, on Excel, or by hand. The formula for the number of combinations of n items taken k at a time is given by:

$$_nC_k = \frac{n!}{(n-k)!k!}$$

Example:

A flight has 3 open seats but there are 7 people on stand-by. How many different ways can the three people be chosen?

Solution:

Since the 3 people will all get a seat, it does not matter what order they are chosen, so this is a combination problem with 7 items taken 3 at a time. We can subtract 3 from 7 to get 4. Then 7! can be broken down to 7*6*5*4!. Notice we stopped at 4! because that will cancel in the denominator leaving the 3! as 3*2*1 and the 7*6*5 in the numerator. Simplifying the numerator and denominator we see that a factor of 6 will cancel leaving the total number of arrangements as 35.

$$_7C_3 = \frac{7!}{(7-3)!3!} = \frac{7!}{(4)!3!} = \frac{7*6*5*4!}{4!*3*2*1} = \frac{7*6*5}{3*2*1} = 35$$

Now we have two methods to use to count the number of arrangements of n items taken k at a time. However, that means we need to be able to correctly choose between the methods.

Example:

Let's consider a race where the top three finishers go on to race in the final round. How many different arrangements of three people can go on to the final round if there are 15 runners in the race?

Solution:

We will first need to determine if this is an application of combinations or permutations. In this case, we need to decide if changing the order gives us the same or a new arrangement. Since changing the order of the finishers will not change the fact that the same three progress to the finals, this is a combinations problem. We are finding the number of combinations of the 15 runners taken 3 at a time.

$$_{15}C_3 = \frac{15!}{(15-3)!3!} = \frac{15!}{(12)!3!} = \frac{15 * 14 * 13 * 12!}{12! * 3 * 2 * 1} = \frac{15 * 14 * 13}{3 * 2 * 1} = 455$$

Here we can first subtract 3 from 15 to get 12 factorial in the denominator so we count down the 15! to 12! So we can cancel 12! From the numerator and denominator, leaving 15*14*13 on top and 3*2*1 on the bottom. We can cancel a factor of 6 leaving 455 different arrangements for the combination of three winners. Of course, if we wanted the winners to get a first, second, and third place ribbon, this would become a problem for permutations because each arrangement would change the ribbons that the runners would receive.

This problem can also be calculated using Excel. The command is =COMBIN(15,3)

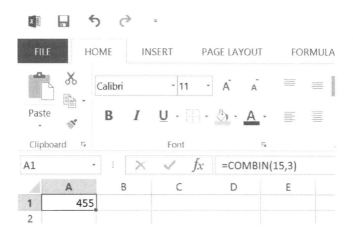

Probability

We will focus on two types of probability in this chapter, Theoretical and Empirical. **Theoretical Probability** is the probability of something happening in theory. That means that it is the number of times something can occur divided by the total number of possible occurrences when the outcomes are equally likely. **Empirical Probability** is the actual rate at which the outcome occurred in past experience. In both cases we have the basic probability expressed as P(x) meaning the probability that x will occur.

Theoretical P(x) = $\dfrac{the\ number\ of\ ways\ x\ can\ occur}{the\ total\ number\ of\ possible\ outcomes}$

Empirical P(x) = $\dfrac{the\ number\ of\ times\ \ x\ did\ occur}{the\ total\ number\ of\ outcomes}$

We can use things like cards or dice to calculate either of these probabilities. For example, let's say a die was rolled 15 times with the following outcomes:

Outcome of roll	1	2	3	4	5	6
Number of outcomes	3	4	2	2	1	3

We could find the probability of rolling a 3 on the die.

$$\text{Theoretical } P(x) = \frac{\textit{the number of ways x can occur}}{\textit{the total number of possible outcomes}} = \frac{1}{6}$$

$$\text{Empirical } P(x) = \frac{\textit{the number of times x did occur}}{\textit{the total number of outcomes}} = \frac{2}{15}$$

Notice that the probabilities are different for the same outcome. For Theoretical probability, there is only one way to roll a 3 and six different possible numbers that could come up. For Empirical probability there were a total of 15 rolls and 2 of them were the number three.

There is actually a "Law of Large Numbers" that says that as the sample size gets larger, the Empirical probability will approach the value of the Theoretical probability.

Theoretical Probability

Some simple examples of Theoretical probability would be determining the probability of selecting a particular card or rolling a particular number on a die. For example the probability of rolling a number less than 3 on a die is the number of outcomes that are less than 3 (two possible outcomes less than 3) divided by the total number of outcomes (6 total numbers on a die) for the probability of P(rolling a number less than 3) = 2/6 = 1/3. The examples can also get more complicated such as examples that use permutations or combinations to determine the total number of outcomes.

Example:

If you selected 3 stocks from a group of 9 that increased in value and 6 that decreased in value this quarter, what is the probability that all three stocks increased in value?

Solution:

To find the probability that all 3 stocks went up in value, we would find the ratio of the number of ways of selecting 3 stocks that went up in value to the number of ways of selecting any 3 stocks. Since the order the stocks are chosen does not matter, it will still be the same 3 stocks, this problem uses combinations to determine the total number of ways to select 3 stocks.

A = selecting 3 stocks that increased in value

P(A) =

$$\frac{Ways\ of\ selecting\ 3\ stocks\ from\ 9\ that\ increased\ in\ value}{Number\ of\ ways\ of\ selecting\ 3\ stocks\ from\ all\ \ stocks}$$

$$P(A) = \frac{{}_9C_3}{{}_{15}C_3}$$

$$P(A) = \frac{\frac{9!}{(9-3)!3!}}{\frac{15!}{(15-3)!3!}} = \frac{\frac{9!}{(6)!3!}}{\frac{15!}{(12)!3!}} = \frac{\frac{9*8*7*6!}{(6)!3!}}{\frac{15*14*13*12!}{(12)!3!}} = \frac{\frac{9*8*7}{3!}}{\frac{15*14*13}{3!}}$$

$$P(A) = \frac{\frac{9*8*7}{3*2*1}}{\frac{15*14*13}{3*2*1}} = \frac{84}{455} = \frac{12}{65}$$

Calculating the combinations using technology will save you a lot of steps and time so you could have gone straight to:

$$\frac{{}_9C_3}{{}_{15}C_3} = \frac{84}{455} = \frac{12}{65}$$

Example:

We can determine theoretical probability for things such as:

a. Find the probability of drawing a heart from a standard deck of cards.
b. Find the probability of having exactly 2 girls out of three children.

Solution:

a. Since there are 52 cards in the standard deck, and four suits. That means 13 of those 52 cards are hearts, so

$$P(\text{heart}) = \frac{13}{52} = \frac{1}{4}$$

b. The possible outcomes of gender for three children are made up by the set of these eight possible outcomes: {ggg, gbb, gbg, bgg, bbg, bgb, gbb, bbb}

$$P(\text{exactly 2 girls}) = \frac{3}{8}$$

Empirical Probability

The second type of probability we will explore is Empirical Probability. It is the probability calculated from previous results, or empirical data. In other words, it is the ratio of the actual number of times a particular outcome occurred compared to the total number of outcomes that occurred.

$$P(x) = \frac{Number\ of\ outcomes\ that\ satisfy\ the\ condition\ of\ x}{Total\ number\ of\ outcomes\ possible}$$

Notice that the equation is the same; it is how we find the number of outcomes that is different.

Example:

	Men	Women
Math major	18	16
Science major	27	39

If someone is in the department of math and sciences, what is the probability they are a math major?

Solution:

Since the data is given, you add up the math majors and divide it by the total number of students in the department to get:

$$P(\text{math major}) = \frac{\textit{number of math majors}}{\textit{number of majors}} =$$

$$\frac{18 + 16}{18 + 16 + 27 + 39} = \frac{34}{100} = \frac{17}{50}$$

Probabilities can be expressed in fraction, decimal, or percentage form.

Chapter Review

Factorial Notation – Notation given by an exclamation mark (!) after a constant. The value is the product of the number and each counting number less than the number given in the factorial.

Fundamental Counting Principle – states that if you have m items of one type and n items of another, there are m*n ways you can combine those items.

Permutations – The number of ways a certain number of objects can be arranged if the order of arrangement is important.

$$_nP_k = \frac{n!}{(n-k)!}$$

Combinations – The number of ways a certain number of objects can be arranged if the order of arrangement is NOT important.

$$_nC_k = \frac{n!}{(n-k)!k!}$$

Theoretical Probability – The probability of something happening in theory

Empirical Probability – The probability of something happening based on the actual rate of occurrence in past experience.

Law of Large Numbers – As the sample size increases, the empirical probability will approach the theoretical probability.

Explorations

Writing Across the Curriculum:

1. Explain in your own words why empirical probabilities are used in determining premiums for life insurance policies.

2. Explain in your own words, how to find the theoretical probability of an event.

3. How are permutations and the fundamental counting principle the same? How are they different?

4. How are Theoretical and Empirical probability the same? How are they different?

5. What is Combinatorics? Who is an influential mathematician in that field and what has he/she done?

6. How might counting principles be used to enhance community?

Critical Thinking and Values for Effective Problem Solving:

1. Two playing cards are dealt to you from a well-shuffled deck of 52 cards. If either card is a diamond, or both are diamonds, you win;

otherwise, you lose. Determine whether this game favors you, is fair, or favors the dealer. Explain your answer. Can you find another game like this?

2. Should some forms of gambling, such as this one, be legalized? Why or why not? Discuss who is often hurt by gambling and who is often helped. Does your response fit with the value of Community?

3. What forms of gambling are legal and what forms are not? Give examples. Should gambling be allowed on Indian reservations in states where it is not allowed off those reservations? Why or why not? How do these activities support or break down community?

Team Exercise:

Can people selected at random distinguish Coke from Pepsi? Design an experiment to determine the empirical probability that a person selected at random can select Coke when given samples of both Coke and Pepsi. Describe the experiment and document your results.

Exploring Number Theory

Fall 2013 Headcount

Total University Enrollment:	16,275
University Campus, Saint Leo, Florida	2,234
Graduate Programs	3,650
Adult Education Center	409

Regional Centers	7,057
Center for Online Learning	3,162
Online Consortium of Independent Colleges and Universities	172

Saint Leo University Facts and Figures 2013-2014

1. Based on the data given, what were the odds that a randomly selected student was from a graduate program at SLU in fall 2013? Based on the data, how many SLU students were fully online in all? Choose one of the specific locations and determine the odds that a randomly selected student came from that location. Make sure you can justify your answers.

2. Choose two or more of the specific locations and determine the probability that a randomly selected student attended either of the locations in fall 2013. For example, find the probability that a randomly selected student attended a regional center or the adult education Center?

3. Choose one of the locations and determine the total number of ways that a committee of 3 students could have been chosen to represent that location.   Choose one of the locations and determine the total number of ways that a committee consisting of a lead representative, second representative and alternate could have been chosen to represent that location.

Chapter 6: Probability Theory

In the last chapter we found the probability an event would occur. In this chapter we will begin by finding the probability an event will not occur. This concept is called the compliment and is expressed as $P(\bar{x})$ and it is given by $P(\bar{x}) = 1 - P(x)$

Example:

When you draw a card from a standard deck, find the probability that the card is not a five

Solution:

Since there are 52 cards in the deck and four of them are the number 5,that means the probability of drawing a 5 is $\frac{4}{52} = \frac{1}{13}$ so the probability that the card is not a five is $1 - \frac{1}{13} = \frac{12}{13}$.

Example:

Using the table below what is:

a. The probability that a man in the Department of Mathematics and Sciences majored in Science?
b. The probability that a man in the Department of Mathematics and Sciences did not major in Science?

	Men	Women
Math major	18	16
Science major	27	39

Solution:

a. $P(x) = \dfrac{number\ of\ men\ who\ majored\ in\ science}{total\ number\ of\ men} =$

$\dfrac{27}{18+27} = \dfrac{27}{45} = \dfrac{3}{5}$

b. $P(\bar{x}) = 1 - P(x) = 1 - \dfrac{3}{5} = \dfrac{2}{5}$

"And" and "Or" Probabilities

We can also find the probability of more than one event occurring. We might be trying to determine if two events happened together, or if one or more of two events happened at all. For two events happening together, we multiply their separate probabilities together or for empirical data we find the overlapping cells in the table of data. For one "or" another probability happening, we add their probabilities in such a way that we do not add any outcome more than one time. If we add totals that contain overlapping cells, then we would need to subtract those overlapping cell values so they are only included once in our total.

Example:

Consider the table below, of students in the Honors Program. We can find several different combined probabilities such as those listed below.

	Freshman	Sophomore	Junior	Senior	Total
Male	66	62	51	35	214
Female	71	68	59	41	239
Total	137	130	110	76	453

    a. Find the probability that an honors student is an upper classman.

    b. Find the probability that an honors student is female or a sophomore.

    c. Find the probability that an honors student is a freshman.

    d. Find the probability that an honors student is not a freshman.

    e. Find the probability that an honors student is a male senior.

    f. Find the probability that an honors student is a senior or male.

Solution:

a. P(upper classman) =

$\dfrac{number\ of\ Juniors\ and\ seniors}{number\ of\ honors\ student} = \dfrac{110+76}{453} = \dfrac{186}{453} = \dfrac{62}{151}$

b. P(female or sophomore) =

$\dfrac{number\ of\ females\ and\ sophomore\ without\ repeating}{number\ of\ honors\ student}$

$= \dfrac{239+130-68}{453} = \dfrac{301}{453}$

or $= \dfrac{71+68+59+41+62}{453} = \dfrac{301}{453}$

c. P(freshman) $= \dfrac{number\ of\ freshman}{number\ of\ honors\ student} = \dfrac{137}{453}$

d. P(not a freshman) = 1 - P(freshman) = $1 - \dfrac{137}{453}$

$= \dfrac{316}{453}$

e. P(male senior) = P(male and senior at the same time) =

$\dfrac{number\ of\ males\ and\ seniors\ at\ the\ same\ time}{number\ of\ honors\ student} = \dfrac{35}{453}$

f. P(senior or male) =

$\dfrac{number\ of\ males\ and\ seniors\ without\ repeating}{number\ of\ honors\ student} =$

$\dfrac{214+76-35}{453} = \dfrac{325}{453}$

Example:

Given a deck of cards:

a. Find the probability of drawing a 3 or a queen.
b. Find the probability of drawing a 2 and a 6.
c. Find the probability of drawing a 2 and a club

d. Find the probability of drawing a king or a heart.

Solution:

a. P(3 or queen) =

$$\frac{Total\ number\ of\ 3s\ and\ queens\ without\ repeating}{Total\ number\ of\ cards} =$$

$$\frac{4+4}{52} = \frac{8}{52} = \frac{2}{13}$$

b. P(2 and 6) =

$$\frac{Total\ of\ cards\ that\ are\ a\ 2\ and\ a\ 6\ at\ the\ same\ time}{Total\ number\ of\ cards} =$$

$$\frac{0}{52}$$

c. P(2 and club) =

$$\frac{Total\ cards\ that\ are\ a\ 2\ and\ a\ club\ at\ the\ same\ time}{Total\ number\ of\ cards} =$$

$$\frac{1}{52}$$

d. P(king or heart) =

$$\frac{Total\ cards\ that\ are\ a\ king\ or\ a\ heart\ without\ repeating}{Total\ number\ of\ cards}$$

$$= \frac{4\ kings + 13\ hearts - 1\ king\ of\ hearts}{52} = \frac{16}{52} = \frac{4}{13}$$

Conditional Probability

We can refine our probability calculations, based on specific categories when conditions are given. For example, if we look at the math and science example above, if we know the person is a math major, we only have to calculate the probability out of all math majors rather than all students. Conditional probability means you first limit the outcomes to only those that meet the condition, then calculate the probability within that restricted pool of values. The notation for conditional probability is P(a | b) or the probability of a given b.

Example:

Use the given data to find the following probabilities.

	Freshman	Sophomore	Junior	Senior	Total
Male	66	62	51	35	214
Female	71	68	59	41	239
Total	137	130	110	76	453

a. The probability that someone is a sophomore given that they are female.

b. The probability that someone is male given that they are a senior.

c. The probability that someone is an upper classman given that they are male.

Solution:

a. P(sophomore | female) =

$$\frac{Total\ number\ of\ females\ who\ are\ sophomores}{Total\ number\ of\ females} = \frac{68}{239}$$

b. P(male | senior) = $\dfrac{Total\ of\ seniors\ who\ are\ males}{Total\ number\ of\ seniors}$

$$= \frac{35}{76}$$

c. P(upper classman | male) =

$$\frac{Total\ number\ of\ males\ who\ are\ upper\ classman}{Total\ number\ of\ males}$$

$$= \frac{51 + 35}{214} = \frac{86}{214} = \frac{43}{107}$$

Odds

A concept very similar to probability is odds. The difference is that odds are the number of outcomes in favor divided by the number of outcomes against, while probability is the number of outcomes in favor divided by the total number of outcomes. The numerators are the same, but the denominator for odds is the difference between the numerator and denominator of the probability. That means if the probability of an outcome is given by $P(x) = \dfrac{a}{b}$ then the

odds are given by odds in favor of $x = \dfrac{a}{b-a}$.

The odds against x would be the reciprocal since it would be the number outcomes against divided by the number of outcomes in favor.

Odds can also be expressed as a: (b-a) or a to (b-a)

Example:

The table below shows the number of people who have each type of pet in a certain survey.

	Number
Dog	64
Cat	31
Bird	5
No pet	30
Total	130

a. Find the odds that a randomly selected person has a dog.
b. Find the odds that a randomly selected person does not have a dog.
c. Find the odds that a randomly selected person has a bird.
d. Find the odds that a randomly selected person has a cat or a bird.

Solution:

a. Odds of a dog $= \dfrac{number\ of\ dogs}{number\ that\ are\ not\ dogs} = \dfrac{64}{31 + 5 + 30}$

$= \dfrac{64}{66} = \dfrac{32}{33}$ or 32:33

b. Odds of not a dog $= \dfrac{1}{odds\ of\ a\ dog} = \dfrac{1}{\frac{32}{33}} = \dfrac{33}{32}$

c. Odds of a bird $= \dfrac{number\ of\ birds}{number\ that\ are\ not\ birds} = \dfrac{5}{130} = \dfrac{1}{26}$

or 1:26 or 1 to 26.

d. Odds of a cat or bird =

$\dfrac{number\ of\ a\ cat\ or\ a\ bird}{number\ that\ are\ not\ cats\ or\ birds} = \dfrac{31 + 5}{64 + 30} = \dfrac{36}{94}$ or 18:47

or 18 to 47.

Expected Value

Many games of chance or bids for a contract are based on expected value. They each have a cost associated with them such as the cost to put together a bid on a project, or the cost to enter a game of chance. They also have a probability of success and a relatively defined pay-out if successful. When you put these concepts together you can determine the expected value, or average outcome if you were to participate repeatedly. If you expect to have a positive outcome in the long run, then you might choose to participate, if you expect to lose in the long run, you should probably pass on the bid or game of chance. Casino games always have a negative expected value because that is how the casino makes money, by the overall average going to the casino exceeding the amount they pay out, even though there are occasional people who win more than they spend.

Example:

For the example below, find the expected value for each contract and determine whether you should bid on one, the other, both, or neither contracts. The bid cost is how much it will cost you to make the bid, Payment is the amount you can expect to make if you are awarded the contract. The probability of hire is the probability you will win the bid.

	Bid Cost	Payment	Probability of hire	Expected value
Contract A	$6000	$170,000	.25	
Contract B	$15000	$460,000	.15	

Solution:

Expected value of Contract A = $170,000*(.25) – $6000 = $36,500

Expected value of Contract B = $460,000*(.15) - $15,000 = $54,000

The expected value of both contracts is positive so they are worthwhile pursuing.

Chapter Review

"And" Probabilities – The probability of two events happening together.

"Or" Probabilities – The probability that either of two events will occur.

Conditional Probability – The probability of an event happening after first limiting the possible outcomes to those that meet a given condition.

Odds – The number of outcomes in favor divided by the number of outcomes against.

Expected Value – The combination of the probability of an occurrence coupled with the expected benefit.

Explorations

In The News: Financial Facts:

Go to a newspaper database and search for "probability". Choose one of the articles from your search results and explain how the term Probability is used in that article. Also explain why the given probability is or is not used correctly in the article.

DON'T FORGET TO REFERENCE ! See *Section 1* for an example of how.

Critical Thinking with Probability:

Almost anyone in the United States can play the lottery and lottery games come in many kinds. You might be surprised how many people dedicate $50 (or more) a week to playing lotto games. But can the game you choose significantly affect you chances of winning? For the next two exercises you will need to use the Internet to find information.

1. Choose a state that runs lotteries.
   a) Calculate your odds of winning if you spend $1 on an entry.
   b) Calculate your odds of winning if you spend $50 on an entry.
   c) Compare these odds to the odds of being in a car accident, plane crash, struck by lightning, or hit by a meteorite. (These numbers *are* out there so search for them!)

2. At the rate of $50/week you will have spent $10,000 in less than 5 years. Take a look at the results of exercise 1 through 3 under Money Talk in the previous section. Given those results and your calculation of the odds, are lotteries a good personal investment?

Discussion:

1. While many states run lotteries, most prohibit individuals and businesses from doing so. Is that ethical? How does this impact the sense of community?

2. Court cases have held bartenders liable for serving alcohol to someone who is already drunk. Gambling is an addiction like alcohol. Should a state be liable for selling lotto tickets to a gambling addict? Do you have the same response when you consider the issue only from the value of community?

Writing Across the Curriculum:

1. Write and explain the formula used to find the expected value of an experiment with two possible outcomes and with three possible outcomes.

2. If the expected value and cost to play are known for a particular game of chance, explain how you

can determine the fair price to pay to play that game of chance. Give the formula for determining the fair price to pay to play a particular game of chance with three possible *gross* amounts that can be won.

3. The dealer shuffles five black cards and five red cards and spreads them out on the table face down. You choose two at random. If both cards are red or both cards are black, you win a dollar. Otherwise, you lose a dollar. Determine whether the game favors you, is fair, or favors the dealer. Explain your answer.

4. If events A and B are mutually exclusive, explain why the formula P(A or B) = P(A) + P(B) – P(A and B) can be simplified to P(A or B) = P(A) + P(B).

5. Explain how to determine probabilities when you are given an odds statement.

Chapter 7 Introduction to Statistics

In this section we will explore the terminology of Statistics including types of data, critical thinking, and the design of experiments.

Statistics is the science of collecting, organizing, summarizing, describing, and interpreting data. In this course we will explore descriptive statistics, where data is used to describe a sample. Another type of statistics is inferential where sample data is used to make predictions about a population

The **population** is the complete collection of all the elements that we are interested in, while the **sample** is any subset of a population. With both of these, a **data point** is the value associated with the property or characteristic of one element of the population. Data may be a number, a word, or a symbol.

An **experiment** is a planned activity performed to collect data. The data might be a **parameter,** a single numerical value summarizing all the data for an entire population. Or it might be a **statistic,** a single numerical value summarizing all the data for a given sample. We will spend most of our time with **quantitative**, or numeric data. However, **categorical** data, or data that fits categories but cannot be organized numerically, is also common. Categorical data can be numeric, such as zip codes or team jersey numbers, but those numeric values do not fit a numeric organization. Data can also be continuous or discrete. **Discrete** data is something you count, like points in a basketball game or the number of fouls, while **continuous** data is something you measure, such as the height of the rim or how tall the players are.

Example:

Population: the set of all students at Saint Leo University

Sample: the set of all students at Saint Leo University enrolled in the Mathematics program

Parameter: the average GPA of all students at Saint Leo University

Statistic: the average GPA of all Mathematics majors at Saint Leo University

Critical Thinking

Whenever you review statistical results, or work with statistics, you need to think critically about several key ideas including source, context, sampling method, conclusions, and significance. Each of these can influence the validity of the study.

The **source** is the person or group responsible for the study. You should determine if they are objective or if there is a personal or professional interest in the outcome of the study that might lead to bias. Is there an incentive to distort the results by either the person conducting the study or the person or group funding the study?

The **context** of the data takes into account the actual data such as where it came from, how and why it was collected. You should make sure you understand why the study was done and how it was done.

The **sampling method** is critical, because no matter how careful you are when analyzing the data; nothing can make the experiment valid if the data was collected in a way that is not representative of the group being studied. Hence, you must know and understand the purpose of the sampling method before accepting any results from a study.

The **conclusions** and **significance** go hand in hand. The conclusion should be precisely related to the study and not broader than the variable that was tested. The findings would also need to be significant enough that the results are statistically expected to occur if the study was repeated again. You can then go one step further to determine whether the results have any practical implications, so that they would also have practical significance rather than just statistical significance.

Frequency Distribution Tables

Frequency distribution tables, graphs, and plots are used to organize, summarize, and describe the data. In this section we will cover frequency tables and various types of graphs and plots.

Data obtained from an experiment can be cumbersome and difficult to comprehend. A frequency distribution table is often used to summarize and organize the data.

A **frequency distribution table** is a table that associates each value or class of values of the data to its frequency, the number of times the value occurs. The two types of frequency distribution tables are grouped and ungrouped frequency distribution tables.

Example of an ungrouped frequency table:

Data	Frequency
10	2
20	6
30	10
40	2

Note that in the above example there are n = 2+6+10+2 = 20 numbers in the data set.

Example of a grouped frequency table:

Data	Frequency
1-10	2
11-20	6
21-30	10
31-40	2

There are several definitions that relate to grouped frequency distributions. The left column in a frequency distribution is referred to as the classes or groups of data while the right side represents the frequencies, or the number of times each data value occurs in each class of the data set.

For each class, there are **lower** and **upper class limits**. The lower class limit is the left side value for each class, while the upper class limit is the right side of each class.

Between classes we have something called **class boundaries**. This is the halfway point between the upper limit of one class and the lower limit of the next class. It is referred to as the upper boundary of one class and the lower boundary of the next class. In order to find the lower boundary of the first class, find the difference between the upper limit and the upper boundary of the first class and then subtract this value from the lower limit of the first class. In order to find the upper boundary of the last class, find the difference between the lower limit and lower boundary of the last class and then add this value to the upper limit of the last class.

Class midpoints are the midpoints for each class and can be found by adding the lower and upper class limits of a class and then dividing by two.

Finally, the **class width** is the distance between any two consecutive lower class limits, or any two consecutive upper class limits.

Example:

Data	Frequency
0-99	5
100-199	8
200-299	13
300-399	12
400-499	19

Lower class limits, 0, 100, 200, 300, 400

Upper class limits: 99, 199, 299, 399, and 499

Class boundaries: -0.5, 99.5, 199.5, 299.5, 399.5, 499.5

Class midpoints: 49.5, 149.5, 249.5, 349.5, 449.5

Class width: 100

Constructing a Frequency Table

When constructing a frequency table, you should look at the size of your data set and select between 5 and 20 classes, remembering that you want to break into only enough groups to recognize patterns in the data. Fifteen to twenty classes are usually only used in cases of extremely large data sets. Then, look at the range of your data set and divide that by the number of classes that you selected. Take your result and **round up** to the next integer to get your class width. If you mistakenly round down, your last class might end before you get to your last data point, although you can always add another class. As long as you include all the data, you can have a few more or less classes than you originally planned, so you should feel free to round slightly higher or slightly lower. This flexibility might allow you to choose a more convenient class width such as 5 rather than 6, or 10 rather than 8.

In starting your first class, you should look for a convenient number that is at or below your first data point. For example, if your data point is either 10, 11, 12, 13, or 14, it might be convenient for you to start your first class at 10 because that is a nice round number to work with. Then, add the class width to get the next lower class limit. Once all the lower class limits have been filled in, you would choose the upper class limit as the closest value to the next lower class limit. For example, if your lower class limits are 10, 20, 30, ... then your upper class limits would be 19, 29, 39, ... While if your lower class limits are 25.0, 30.0, 35.0, ... Then your upper class limits might be 29.9, 34.9, 39.9, ...   Continue until your classes are determined.

Now you are ready to fill in the frequency for each class. The class frequency for a class is the total number of data points that fall within the boundaries of that class. You find this number by counting the data points within each range and then record that number on the right side column of the frequency table.

Example:

Given the data below, construct a frequency table.

6, 8, 9, 11, 7, 3, 3, 5, 9, 10, 12, 0, 4, 10, 7, 6, 8, 9, 4,
5, 13, 1, 4, 4, 5, 3, 6, 9, 2, 3, 4, 8, 10, 9, 7, 6, 2, 6, 11,
1, 4, 5, 8, 7, 6, 3, 8, 9, 10, 7, 7, 7, 10, 6, 8

Solution:

There are 55 data points that range from 0 to 13. This is a rather small data set so I will go with 5 classes. Furthermore, since 15 is divisible by 5 classes, I will go from 0 to 14 (15 units) and with 5 classes that makes each class 3 units wide. My lower class limits will be 0, 3, 6, 9, and 12; so my upper class limits must be 2, 5, 8, 11, and 14.

Frequency table:

Data	Frequency
0-2	5
3-5	15
6-8	20
9-11	13
12-14	2
Total	55

Organizing Data Using Graphs and Plots

We will review several ways to display data graphically including:

- Histograms,
- Frequency Histograms,
- Dot Plots,
- Scatter Diagrams,
- Pie Charts,
- Stem and Leaf Plots.

Histograms

A histogram can illustrate trends in large amounts of data. A histogram is a bar graph in which the height of each bar represents the frequency of each class of data. Similarly in a relative frequency histogram the height of each bar represents the relative frequency of each class.

In histograms, you can use either the class limits or the class boundaries to label the classes along the horizontal axis.

Example:

Construct a histogram for the following frequency distribution table.

Data	Frequency
0-2	5

3-5	15
6-8	20
9-11	13
12-14	2

There are 5 classes, so there will be 5 bars in the histogram. Notice there are no breaks between the bars in a histogram because this graph can be used for continuous as well as discrete data. A bar graph can replace a histogram for discrete data since breaks between the bars are fine in that situation. Technically, a histogram will not have any breaks between bars. If the graph has breaks between bars, then it is a bar graph.

Frequency Histogram

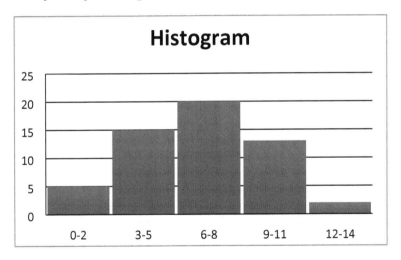

Dot Plots

A dot plot is useful to display the grouping of a limited amount of data. In a dot plot each value of data is represented by a dot on a number line above the value that corresponds to the data point. If there is already a dot, then the new dot goes above the dot that already exists for that value. Common values appear as higher areas of the dot plot while less common areas are lower or have no dots at all.

The dot plot is one of only a few types of graphs that retain the original value of every data point so that the plot could be used to reconstruct the original set of data. In most graphs the data is first organized into a grouped frequency distribution and then graphed so only the frequency of classes can be determined, not the original data values.

Example:

Construct a dot plot for the following frequency distribution table.

Data	Frequency
1	1
2	4
3	3
4	2

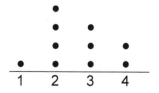

Pie Charts

A pie chart is a great way to display the relative percentages of the values within a data set. In a pie chart each slice represents a class of data. The relative frequency distribution is often used for a Pie Chart because each slice of the pie represents the proportion of data in that class. For example, if one class contains 25% of the data, then the corresponding slice of pie will be ¼ or 25% of the pie.

Example:

Construct a Pie chart of the following data.

Data	Frequency	Relative Frequency
0-2	5	9.09%
3-5	15	27.27%
6-9	20	36.36%
9-11	13	23.64%
12-14	2	3.64%

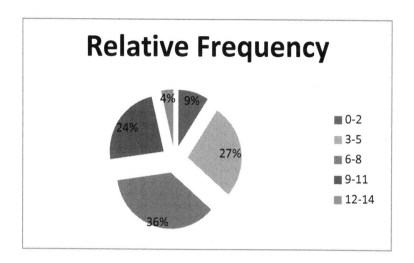

Scatter Diagrams

A scatter diagram shows the dispersion of a data set and is helpful when trying to determine general trends in the data. A scatter diagram is a plot of the paired (x, y) data with a horizontal x-axis and a vertical y-axis. This is a common type of plot used in regression analysis where we try to find the trend in the data. This plot is usually used whenever data is arranged in ordered pairs.

Stem-and-leaf Plots

In a stem-leaf plot each data is represented by two parts: stem and leaf. The stem is often the first digit of each number while the leaf is made up of the remaining digits. The stem is listed once on the left side of the plot and then each number is expressed individual on the right side of the plot as the remaining leaf. The stem may occasionally contain more than one digit and the leaves are the remaining digits.

Example:

Data: 25, 26, 29, 30, 30, 35, 38, 45, 48, 48, 49

Stem and leaf plot:

Stem	Leaves
2	5 6 9
3	0 0 5 8
4	5 8 8 9

Graphing Summary

The dot plot and stem and leaf plot are often used with small sets of data or especially when you want to retain the value of the original data. Histograms and Pie charts are a way to summarize larger sets of data when the individual points do not need to be retained.

When you are looking at scatter plots and histogram graphs always check to be sure that the graphs start with a vertical axis value of zero. Sometimes a researcher will only include the top portion of the graph, starting their bar chart far above zero, in order to emphasize a difference that appears at the top of the graph. Other researchers might use pictures or colors to manipulate your emotions rather than focusing on the meaning of the data with the graph. Always look at what the graph is representing before allowing a researcher to use your emotions or reflexes to come to a conclusion that they intend but that may not be supported by the true nature of the data.

Center, Variation, and Relative Standing

Measures of center, variation, and relative standing (position) are used to gain valuable insight about important characteristics of data. In this chapter we shall investigate the major concepts associated with the measures mentioned above. The topics covered are as follows:

Measures of center:
- Mean
- Median
- Mode
- Midrange

Measures of variation:

- Range
- Standard Deviation
- Variance

Measures of relative standing (position): Z-score, quartiles, percentiles Chebyshev's Rule, and the Empirical Rule, the Range Rule of Thumb.

Measures of Center

Measures of center are numerical values that provide us with information about the center of the data. These are mean, median, mode, and midrange.

Mean

The most important measure of center is mean of the data, which is denoted by $\bar{x}$ (x-bar). Mean of a set of data is simply the weighted average of the data. When the weights of all data points are the same, then the mean is the same as the average.

$\bar{x} = \dfrac{\sum x}{n}$ = (sum of all the numbers in the data)/(number of values in the data)

Example:

The scores in a Statistics quiz are: 97, 83, 48, 73, 64, 91, 84, 85, 88, and 22. Find the mean of the quiz.

Solution:

$$\bar{x} = \frac{\sum x}{n} = (97+83+48+73+64+91+84+85+88+22)/10$$

$$\bar{x} = 735/10 = 73.5$$

The mean is the most common measure of center. It takes all the values of data and their weights into account. It is used in many areas of statistics. It can be pulled by an outlier, but not as heavily as some other measures of center since it utilizes all of the data to counter balance any one extreme value.

Median

The **median** $\tilde{x}$ (x-tilde) is the value of data that is located in the middle position, once the data is arranged in either ascending or descending order. The middle location is determined differently depending upon whether n, the total number of data points, is even or odd. If n is odd, then the median is the value in the center, once the data is ordered. If n is even, then the median is the average of the two values located at the center, once the data is ordered.

Example (n is odd):

Find the median of the quiz scores 97, 83, 48, 73, 64, 91, 84, 85, and 88.

Solution:

Sort the data: 97, 91, 88, 85, 84, 83, 73, 64, 48

Median = 84

Example (n is even):

Find the median of the quiz scores 97, 83, 48, 73, 64, 91, 84, 85, 88, and 22.

Solution:

Sort the data: 97, 91, 88, 85, 84, 83, 73, 64, 48, 22

Median = (84+83)/2 = 83.5

The median is the second most common measure of center. The median is not affected by outliers at all, so it is often used in situations with extreme values such as average home prices or average salaries.

Mode

Mode is the value of the data that appears with the highest frequency.

Example:

Data: 25, 55, 75, 75, 75, 85, 95

Solution:

Mode: 75 because it appears 3 times in the data

Example:

Data: 25, 55, 55, 55, 75, 75, 75, 85, 95

Solution:

Modes: 55 and 75

Example:

Data: 25, 55, 65, 75, 85, 95

Solution:

Modes: None, because no single value appears more than once.

Bimodal: If two values of the data share the same greatest frequency, the set of data is said to be bimodal and two modes would be listed.

Multimodal: If more than two values of the data share the same greatest frequency, then the set of data is said to be multimodal and no mode is listed.

Mode is only used as a measure of center either in conjunction with other measures of center, or when the most common data point is the objective.

Midrange

Midrange of a set of data is defined to be the average of the highest and lowest value, or the value half way between the highest and lowest data point.

(highest value of data+ lowest value of data) / 2.

The midrange is only used as a measure of center when the information about a set of data is limited or a very quick estimate is needed. The midrange is often a bias estimate of the center because it is heavily pulled in the direction of outliers since it is calculated using only the most extreme high and low values.

Example:

Find the midrange of the Statistics quiz with scores: 97, 83, 48, 73, 64, 91, 84, 85, 88, and 22.

Solution:

Midrange = (97+22)/2 = 119/2 = 59.5

Note that the midrange is well off from the mean=73.5.

Round-off Rule

A widely used round-off rule in statistics is to carry the computations one decimal place further than the number of decimal places present in the original set of data.

Mean of a Frequency Table

You should find that Statistics at this level is a very consistent science. For example, the round off rule above will repeat each time we apply the mean in other areas such as probability distributions. Remember, the mean is simply the average of all values in the set of data. In the frequency distribution example below, the first line means there are 5 distinct values of 1 and 15 distinct values of 4 within the set of data. That implies that we can find the sum of all 5 values of 1 by multiplying the 5 times the 1. The same can be done by multiplying the 15 by 4 to find the value of all 15 of the data points of 4. The steps are shown below. Remember, if you add up the total frequency, that will give you the total number of values in your data set for your value of n.

Example:

Find the mean of the following table.

Data	Frequency
1	5
4	15
7	20
10	13
13	2
Total	55

Solution:

Mean = Sum(x) / n

Mean = [(1)(5)+(4)(15)+(7)(20)+(10)(13)+(13)(2)]/55

Mean = 361/55 or approximately 6.56

Example:

Find the mean of the following table.

Data	Frequency
0-2	5
3-5	15
6-8	20
9-11	13
12-14	2

Solution:

In this case, we do not know the actual values of the data points, only the range that each value falls within. As a result, the midpoint of each class is the best estimate of the values within that class. That means you should first convert the table to an ungrouped table by replacing each class of data by its class midpoint. In this example, the result will be identical to the table given in the previous example.

Mean = Sum(x) / n

Mean = [(1)(5)+(4)(15)+(7)(20)+(10)(13)+(13)(2)]/55

Mean = 361/55 or approximately 6.56

In order to calculate the mean of a grouped frequency distribution table, you first need to convert the grouped table to an ungrouped table by replacing each class of data by the midpoint of the class. The result will be an approximation.

Measures of Variation

Consider the following sets of data.

Data set #1: 0, 10, 50, 90, 100

Data set #2: 45, 45, 50, 55, 55

The mean and the median for both sets of data is 50. Therefore, if we only take measures of center into account, there is no significant difference between the two data sets. However, it is obvious the two sets are very different. The numbers in the first set are spread out while the numbers in the second set are clustered in the middle. Measures of variation take spread of the data into consideration and provide us with more information about the data. In the next few pages, we will discuss measures of variation: range, variance, and standard deviation.

Range

The range of a set of data is defined to be the difference between the highest and lowest value in the data set

Range = highest value – lowest value.

This measure of spread is not used very often for the same reason the midrange is not popular, it is heavily skewed by extreme values.

Example:

What is the range of the set: 45, 45, 50, 55, 55?

Solution:

The highest value is 55 and the lowest value is 45 so the range is 55-45=10.

Variance and Standard Deviation

We want to come up with a way to measure the spread of the data. To achieve this task, we must somehow include the deviation of each value x from the mean or $x - \bar{x}$ in our calculations. An obvious attempt would be to calculate the average of all the deviations from the mean or $\dfrac{\sum (x - \bar{x})}{n}$. Unfortunately the numerator, the sum of all the deviations $\sum (x - \bar{x})$ is always zero. This is because the mean is the average of all values so they are equally spread above and below the mean so the positive and negative differences cancel each other out. In order to avoid a sum of zero, we will square all the deviations and then find the mean by dividing the sum by n-1. The result is called the **variance of the sample.**

Therefore for a given sample, we have:

Variance of sample = $s^2 = \dfrac{\sum(x-\bar{x})^2}{n-1}$

Standard deviation of sample = $s = \sqrt{\dfrac{\sum(x-\bar{x})^2}{n-1}}$

If the mean is computed by dividing by n, the results are referred to as the variance and the standard deviation of the population. Thus for a given population:

Variance of population = $\sigma^2 = \dfrac{\sum(x-\bar{x})^2}{n}$

Standard deviation of population = $\sigma = \sqrt{\dfrac{\sum(x-\bar{x})^2}{n}}$

In life, we hardly ever have access to all the numbers in the population. Unless it is specifically stated that the numbers represent the entire population, you must use the sample formulas to obtain the variance and the standard deviation.

Example:

Find the variance and the standard deviation of the sample data 45, 45, 50, 55, 55.

Solution:

Obviously n = 5 and mean = (45+45+50+55+55)/5 = 50

If we carry out the problem the long way and organize the information in a table. We strongly recommend that students should use technology to compute the standard deviation.

x	$x - \bar{x}$	$(x - \bar{x})^2$
45	-5	25
45	-5	25
50	0	0
55	5	25
55	5	25

$$\sum (x - \bar{x})^2 = 100$$

Variance = $s^2 = \dfrac{\sum (x - \bar{x})^2}{n-1}$ = 100 / 4 = 25

Standard deviation = $s = \sqrt{25} = 5$

This is an excellent time to get familiar with some form of technology if you have not already. It is impossible to calculate something as simple as standard deviation with some of the sources of data that contain millions of values these days.

Helpful technology might include Excel – using the "insert – function – statistical" menu options. Most financial and graphing calculators also have the basic statistical features we have reviewed so far.

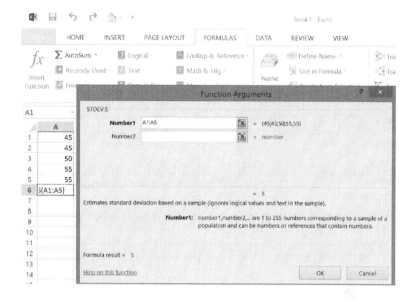

Here I used Excel – using the "insert – function – statistical" menu options. I selected the function STDEV.S for the standard deviation of the sample. The variance is the square of the standard deviation, so rather than do another command, you can just square the result of 5 for the standard deviation to get 25 for the variance.

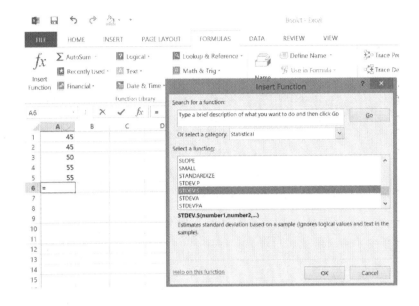

Variance = $s^2 = 5^2 = 25$

Example:

Find the variance and the standard deviation of the sample data 0, 10, 50, 90, 100.

Solution:

Obviously n = 5 and mean = (0+10+50+90+100)/5 = 50

x	$x - \bar{x}$	$(x - \bar{x})^2$
0	-50	2500
10	-40	1600
50	0	0
90	40	1600
100	50	2500
	$\sum (x - \bar{x})^2$ = 8200	

Variance = $s^2 = \dfrac{\sum (x - \bar{x})^2}{n-1}$ = 8200 / 4 = 2050

Standard deviation = $\sqrt{2050} \propto 45.28$

As expected, the standard deviation of this data set is much larger than the standard deviation for the previous example.

Chapter Review

Definitions

Statistics – the science of colleting, organizing, summarizing, describing, and interpreting data

Population – the complete collection of all the elements

Sample – a subset of the population

Data Point – the value associated with the property or characteristic of one element of the population.

Experiment – A planned activity performed to collect data.

Parameter – a single numerical value summarizing all the data for an entire population.

Statistic – a single value summarizing the data for a sample.

Quantitative – data that is numeric

Categorical data – data that fits into categories, but is not numerically meaningful

Discrete – data that we can count

Continuous – data that we can measure

Source – the person or group that is responsible for the study that developed the data

Context – taking into account where the data comes from, how and why it was collected.

Sampling Method – the methods and procedures used when collecting the data sample.

Conclusions – the decision that can be inferred from the data, precisely related to the study and not broader than what was tested.

Significance – a result that is expected to be found if the study is repeated.

Frequency Distribution Table – A table that associates each class of values with its frequency, the number of times it occurs.

Class Limits – The upper and lower limits of the range of value for an entry in a frequency distribution table.

Class boundaries – The dividing line between different classes.

Class Midpoints- The center point of a class of data.

Class Width – The width of a class, the range of values grouped into a single class.

Formulas

		Sample	Population
Mean	$\bar{x}$	$\bar{x} = \dfrac{\sum x}{n}$	μ
Variance	s^2	$s^2 = \dfrac{\sum (x - \bar{x})^2}{n-1}$	σ^2
Standard Deviation	s	$s = \sqrt{\dfrac{\sum (x - \bar{x})^2}{n-1}}$	σ

Explorations

Team Exercises in the News:

1. Find an article that contains at least one statistical graph. Identify the original data as closely as possible from the information given in the study. Do this again for another study that has a different type of graph. In which study were you able to most closely determine the original data? Why? (if you cannot determine much about the original data, then you should find another study to work with)

2. Find a graph or set of data in an article. Each team member should create a different type of graph from the same information. Do the graphs appear to express the same results or give a different impression? Which graph is a good choice and why? Which graph is a bad choice and why?

3. Review news sources to find two different studies that use two different sampling methods such as random sampling, systematic sampling, stratified sampling, or cluster sampling. For both studies, identify the type of study (cross sectional, prospective, or retrospective), type of sampling, population and sample, type of data (qualitative or quantitative and discrete or continuous), level of measurement of the data (nominal, ordinal, interval, or ratio), source of the study, and conclusions of the study. Finally, summarize whether people should or should not accept the findings in the study and explain why.

4. Choose a question of interest with an answer that is unknown to your team members. Work together to construct a study that could be used to attempt to answer your question. Be sure to indicate the type of study (cross sectional, prospective, or retrospective), type of sampling method (random, systematic, stratified, or cluster), population and sample, parameter and statistic, type of data (qualitative or quantitative and discrete or continuous), and the level of measurement of the data (nominal, ordinal, interval, or ratio). Note that you will not conduct the study, just set up the structure for the study based on the concepts in this chapter.

Values and Critical Thinking for Problem Solving:

1. Why is it important that a sample be representative of the population it came from? Why would a bias sample not support the value of community?

2. Use any news source to find an example of a study or graph that is deceptive and explain why you believe it is deceptive and what should have been done differently.

3. If a study showed that many Business majors drop out of school immediately after taking a Statistics class, what is wrong with the following statement? "A study indicates that the course requirement to take Statistics causes many Business majors to

drop out of school." What are other possible lines of reasoning that might be true rather than Statistics causing people to drop out of school? What might be a correct interpretation of the study? Find an example of causation that appears it should be expressed as correlation in a published source.

4. Make up or find a set of data. Graph the set of data two ways that project two different meanings. Include a brief description of how the two graphs imply different meaning. How can the value of community be important when choosing which graph should be used to publicize the information?

5. Find two different graphs for the same topic and explain how and why they are similar and different.

6. Find two different visual representations of the same information and explain why they are accurate or not accurate and how you might be able to express the same information in two different yet equally accurate ways.

Writing across the Curriculum:

1. Write a short essay that can be used to teach someone how to determine the level of measurement of a given data set.

2. Choose any topic, such as basketball. Give an example of something related to that topic that could be studied as a cross sectional study such

as determining the rankings for highest free-throw percentages. Give an example of a retrospective study related to that topic such as how the number of concussions in basketball over the last 5 years compares to other sports. Finally, give an example of a prospective study on the same topic such as following a group of 100 students who played basketball in high school to compare their bone density in their 20s, 30s, 40s, and 50s to those who did not play any sports.

3. Write a short summary that describes pitfalls to be aware of when reading statistical studies. Be sure to include examples.

4. Explain why Integrity and Community are important when reporting statistical findings.

5. Write a short summary that describes techniques to be aware of when interpreting statistical graphs so you are not fooled with bias graphs.

6. Find at least one study about something that interests you and explain what you learned from the study or studies.

Social Justice across the Curriculum:

For each of these questions, consider how the value of Community relates to the situation and the importance of Community when conducting a statistical study.

1. Our national debt can be found in many different sources with the statistics represented many different ways. Identify three or more of these sources and discuss whether any of them provide a plan for the future generations to be able to pay for our spending.

2. What are your views on the country spending more money than it takes in? Support your views with statistical reports or documentation.

3. An "incentive" is often used to encourage people to participate in a study, such as free tickets to an event, money donated to a charity, a chance to win a prize, or even cash. How can an incentive create bias within the study? What group might likely not be included in a study and what group might be over represented when an incentive is used?

4. Find a study that you should be cautious in interpreting because a group sponsors it or organization that you believe might be biased on the issue being presented. Explain whether any bias is apparent within the study.

5. Find a study that can be interpreted in a way that could be harmful to a group in our society (this should be very difficult to find since one rule for any regulated research is that it will do no harm to animals or people)

6. Should authors be held accountable for bias reports? Why or why not? How?

Chapter 8 Statistics

Data is all around us. What does it mean? How can we use it? In this chapter we will start with questions. From there, we will find data that relates to the issue. Finally, we will organize and summarize the data in order to support a conclusion.

For any question, you need to find or collect appropriate data that relates to the issue and is representative of the population of interest. Finding or collecting the data can be more difficult than it seems. Collecting it can be very expensive and time consuming. Finding raw data can also be difficult because there is not a specific set of search words or phrases that will lead to the data. One phrase that worked well for me was "datasets for statistics projects." However, these data sets might be created for educational purposes rather than legitimate data sets collected using appropriate techniques. Government sites such as the National Center for Health Statistics at http://www.cdc.gov/nchs/ have a large amount of data, but it is sometimes difficult to get to the specific information you need. Some educational sites have convenient, organized data sets such as:

- http://college.cengage.com/mathematics/brase/und erstandable_statistics/7e/students/datasets/
- http://panthermath.weebly.com/statistics-ap-data-sets.html,
- http://www2.stetson.edu/~jrasp/data.htm, or

- http://mathforum.org/workshops/sum96/data.collecti
ons/datalibrary/data.set6.html .

Often if you can find a research paper on the topic of interest, the data set might be either referenced or included.

Once you have the question and the data, then it is time to use the techniques from the last chapter to organize and summarize the data.

Example:

Do Pitchers or outfields make more money?

Analysis:

To attempt to answer this question, I would need to find a data set that includes salaries for professional baseball players. If I can find a data set that is already organized into an Excel file or other sortable form, I can go directly to organizing and summarizing. If I find raw data points, then I would first have to enter them into a file.

I am going to use data from Spotrac, at www.spotrac.com, to answer this question. I am able to sort on the site by MLB and then by position.

Now that I have the data, I will need to put all the like positions in the same column so I can sort each column in ascending order by salary in order to create a frequency distribution table.

When I sort the data, the smallest salary for both categories is $507,500. The highest salary for outfield is $25,000,000 while the highest salary for a pitcher is $30,714,286. For my outfielder's frequency distribution table, I am going to classify the data into $3,000,000 blocks. That means the first class will be from 0-$2,999,999 and the last class will run from $24,000,000-$26,999,999 to get the following table:

Outfielders Salary	frequency
0-$2,999,999	78
$3,000,000-$5,999,999	26
$6,000,000-$8,999,999	16
$9,000,000-$11,999,999	10
$12,000,00-$14,999,999	7
$15,000,000-$17,999,999	4
$18,000,000-$20,999,999	5
$21,000,000-$23,999,999	2
$24,000,000-$26,999,999	3
total	151

To organize the Pitchers, I will use the same classes, but have two additional higher classes. Having the same classes will allow me to put my data together to form a double bar graph for a visual comparison.

Salary	Outfielders	Pitchers
0-$2,999,999	78	300
$3,000,000-$5,999,999	26	69
$6,000,000-$8,999,999	16	34
$9,000,000-$11,999,999	10	23
$12,000,00-$14,999,999	7	15
$15,000,000-$17,999,999	4	6
$18,000,000-$20,999,999	5	4
$21,000,000-$23,999,999	2	2
$24,000,000-$26,999,999	3	7
$27,000,000-$29,999,999	0	0
$30,000,000-$32,999,999	0	2
total	151	462

A double bar graph of the totals is not very helpful because there are so many more pitchers, so the totals for each category are almost all much higher. In order to compare the quantities at each pay level, I will convert this to a relative frequency table so the values are in percentage of players for each category.

Salary	Outfielders	Pitchers
0-$2,999,999	0.5165563	0.649351
$3,000,000-$5,999,999	0.1721854	0.149351
$6,000,000-$8,999,999	0.1059603	0.073434
$9,000,000-$11,999,999	0.0662252	0.049784
$12,000,00-$14,999,999	0.0463576	0.032468
$15,000,000-$17,999,999	0.0264901	0.012987
$18,000,000-$20,999,999	0.0331126	0.008658
$21,000,000-$23,999,999	0.013245	0.004329
$24,000,000-$26,999,999	0.0198675	0.015152
$27,000,000-$29,999,999	0	0
$30,000,000-$32,999,999	0	0.004329
total	1	0.999841

Notice that due to rounding the pitcher total is just under 100%, but it is 1 if rounded to the first, second, or third place value. As I glance through the totals, I can observe that a larger proportion of pitchers are at the low end of the scale, but also a larger proportion are at the top. In order to compare all categories at the same time, I will need something graphical so a double bar graph might be effective.

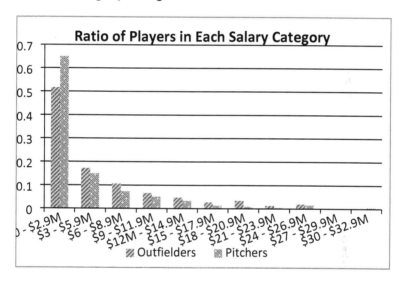

From the bar chart I can see that a larger proportion of pitchers are in the lowest pay category, while a larger proportion of outfielders are in all remaining pay categories that I can distinguish from the graph. That means overall the pitchers are the lowest, but then there are a very few pitchers at the highest end of the pay scale, beyond any outfielders. The data is still not strong enough to conclude which is paid the most.

Summary statistics can help distinguish the middle values, or central tendencies of the data. They can also show the dispersion, or spread of the data.

```
Descriptive Statistics
Column 1

Sample Size, n: 151
Mean:              5.179124e+6
Median:            2.600000e+6
Midrange:          1.275375e+7
RMS:               7.916719e+6
Variance, s^2:     3.609012e+13
St. Dev., s:       6.007505e+6
Mean Abs Dev:      4.613281e+6
Range:             2.449250e+7
Coeff. Of Var.     115.99%
```

The results for outfielders indicate a mean of $5,179,124 and a median of $2,600,000. For the mode I have to look back at my sorted data to see that $507,500 appears the most. The mode is not really relevant since most teams can afford more of the least expensive players, so the mean and median will be the focus. Of course the spread of the data is also important, so I will keep in mind that the standard deviation is $6,007,505, the variance is $36,090,120,000,000 and the range is $24,492,500. Since I will have the standard deviation and range of both sets, I can ignore the variance, as it is the square of the standard deviation.

The summary, or descriptive statistics for the pitcher are also needed.

```
Descriptive Statistics
Column 2

Sample Size, n:  462
Mean:            3.688424e+6
Median:          1.181250e+6
Midrange:        1.561089e+7
RMS:             6.392822e+6
Variance, s^2:   2.732285e+13
St. Dev., s:     5.227126e+6
Mean Abs Dev:    3.655949e+6
Range:           3.020679e+7
Coeff. Of Var.   141.72%
```

The mean salary for pitchers is $3,688,424 with a median of $1,181,250. Here the standard deviation is $5,227,126 and the range is $30,206,790. Notice that both primary measures of center are lower for pitchers than for outfielders. The standard deviation is slightly lower for pitchers, meaning that the salaries are slightly less spread out other than the outliers at the top. As a result, we can conclude that overall pitchers make less than outfield players based on the current statistics. However, a few top pitchers do have a higher salary than the top outfielders.

A pie chart is sometimes a good way to show just how small the number of players at the top salary range is as compared to the lower levels. Just like with the bar graph, the relative frequency or frequency distribution table should be used to form the graph so that the data is in categories. Otherwise, we would have one slice of pie for each different salary so it would just look like a bunch of thin slices and no distinctions would be possible about where a majority of the salaries fall.

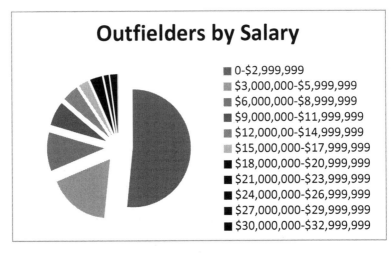

Outfielders by Salary

- 0-$2,999,999
- $3,000,000-$5,999,999
- $6,000,000-$8,999,999
- $9,000,000-$11,999,999
- $12,000,00-$14,999,999
- $15,000,000-$17,999,999
- $18,000,000-$20,999,999
- $21,000,000-$23,999,999
- $24,000,000-$26,999,999
- $27,000,000-$29,999,999
- $30,000,000-$32,999,999

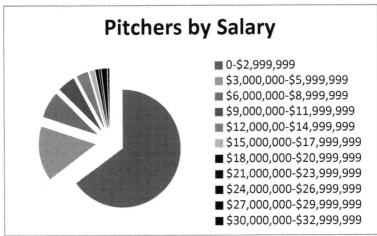

Pitchers by Salary

- 0-$2,999,999
- $3,000,000-$5,999,999
- $6,000,000-$8,999,999
- $9,000,000-$11,999,999
- $12,000,00-$14,999,999
- $15,000,000-$17,999,999
- $18,000,000-$20,999,999
- $21,000,000-$23,999,999
- $24,000,000-$26,999,999
- $27,000,000-$29,999,999
- $30,000,000-$32,999,999

From these pie charts we can see that a much larger portion of pitchers earn lower salaries than outfielders.

Example:

Do men with a higher body fat percentage have larger forearms?

Analysis:

To explore this question a data set is needed that contains body fat percentage and forearm size for the same men. One such set can be found at http://www2.stetson.edu/~jrasp/data.htm

In order to organize the data set, I will need to divide the body fat into classes so I can find the average forearm size for each class. The body fat percentage ranges from 0% to 45.1% so I am going to choose a class width of 6%, resulting in 8 classes.

body fat	Average forearm size
0-5.99	27.3
6-11.99	27.6
12-17.99	28.4
18-23.99	29.0
24-29.99	29.1
30-35.99	30.2
36-41.99	30.5
42-47.99	29.1

Looking at the table, it appears that the forearm size increases as the body fat percentage increases. However, the last data point has a smaller forearm size. We can look at this as a bar chart for a quick visual display that shows the steady increase in forearm size as body fat increases.

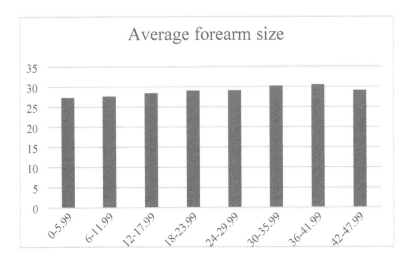

This example seems to be pretty straight forward, so I might want to choose a different question in order to illustrate more statistical features of the data. For example, I might split the data set in half by the top half of the body fat percentage and the bottom half and compare the two groups, like the next example.

Example:

Do taller men have a bigger abdomen?

Analysis:

This analysis can run similar to the previous one, or for something different, the data set can be split in half for the shortest half and tallest half of the subjects, and then analyzed against each other. I will use the same set of data as the previous example.

There are 252 data points, ranging from 29.5 inches to 77.75 inches so a class width of 1 inch can be used. A class width of 2 inches would also be appropriate.

height	average abdomen size
64-64.99	92.3
65-65.99	96.2
66-66.99	89.7
67-67.99	92.6
68-68.99	93.0
69-69.99	69.41
70-70.99	70.33
71-71.99	91.0
72-72.99	91.9
73-73.99	88.4
74-74.99	93.3
75-75.99	91.7
76-76.99	86.5
77-77.99	93.4

Investigating the table there does not appear to be a clear pattern. The largest height is not the largest abdomen size; in fact it is almost the same size as the smallest height.

Looking at a bar graph we can see that there does not seem to be a steadily increasing pattern.

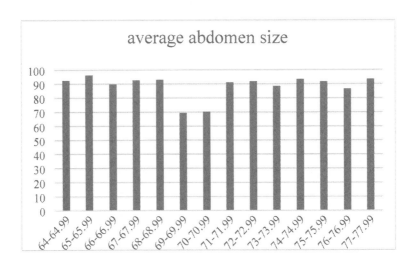

Sometimes this is clearer to see in a line chart:

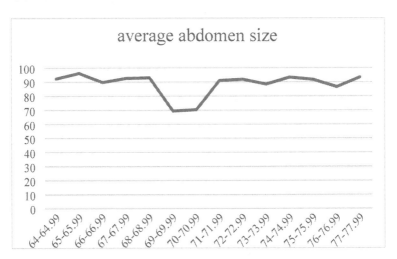

There appears to be a dip in the middle, but otherwise the abdomen size holds pretty steady for most heights.

A numeric comparison is another way to look at the data. There are 252 data points so we can compare the central tendencies and dispersions of the bottom 126 and the top 126 data points.

	shortest	tallest
mean	92.75	92.36
median	91.95	90.8
standard deviation	10.74	10.86
range	52.7	77.7

With the comparison we see that the mean and median abdomen size for the taller men is slightly less than for the shorter men, but also slightly more spread out. There does not seem to be any reason to assume that taller men will have a larger abdomen.

The first example allowed the use of frequency distributions to compare two categories while the last two questions required the mean to be used to compare categories. Our next example finishes with another comparison where relative frequency can be used to compare genders.

Example:

Do men and women have the same heart rate?

Analysis:

A data set with gender and heart rate is needed for this question. A set of data can be retrieved that originated with *A Critical Appraisal of 98.6°F, the Upper Limit of the Normal Body Temperature, and*

Other Legacies of Carl Reinhold August Wunderlich by P. A. Mackowiak, MD, et. al. in *JAMA.* 1992;268(12):1578-1580. The data set has 65 men and 65 women organized into a frequency distribution here.

heartrate	men	women
55-59	1	3
60-64	4	7
65-69	10	11
70-74	24	10
75-79	15	17
80-84	10	12
85-89	1	5
total	65	65

Since there is the same total number of males and females, we do not need a relative frequency distribution table here because we can directly relate the total for each class. It would appear that more woman than men have the lower heart rate categories and also have the higher heart rate categories while men are the more consistent with the majority having a heart rate near the average. We can look at this graphically through a double bar graph to compare bar lengths and also pie charts to look at distribution that way.

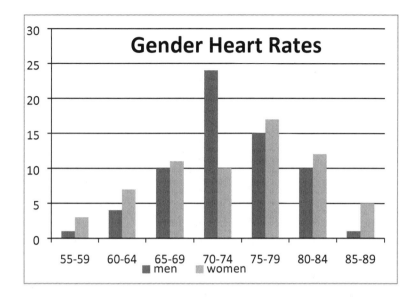

The bar graph shows that heart rates for both genders are relatively normally distributed, with the male heart rates fitting that pattern more closely. All categories are very close except the heart rates from 70-74 where the men significantly outnumber the women.

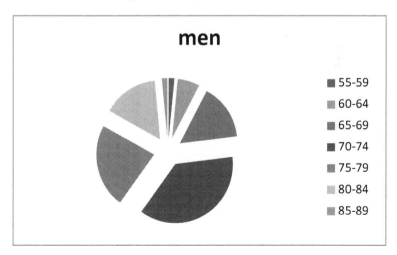

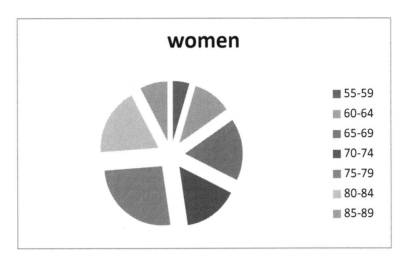

women

- 55-59
- 60-64
- 65-69
- 70-74
- 75-79
- 80-84
- 85-89

The pie chart shows that women's heart rates are much more equally distributed while men tend to fall into two primary categories with very few at the ends.

This still does not answer the question of who has the higher heart rate since women have both the higher and lower rate. We can explore the descriptive statistics of central tendency and dispersion to see if that helps with any conclusions.

	men	women
mean	73.4	74.2
median	73	76
mode	78	79
midrange	72	73
range	28	32
variance	34.5	65.7
standard deviation	5.9	8.1

The descriptive statistics give a much clearer picture that men on average have the lower heart rate. Every measure of central tendency is lower and the dispersion is smaller, indicating that the male heart rates are clustered more closely around the lower values. As a result, we can conclude that although some women have lower heart rates than some men, men have lower heart rates over all.

Explorations

Team Exercise in the News:

1. Choose a question of interest with an answer that is unknown to your team members. Work together to construct a study that could be used to attempt to answer your question. Be sure to indicate the type of study (cross sectional, prospective, or retrospective), type of sampling method (random, systematic, stratified, or cluster), population and sample, parameter and statistic, type of data (qualitative or quantitative and discrete or continuous), and the level of measurement of the data (nominal, ordinal, interval, or ratio). Note that you will not conduct the study, just set up the structure for the study based on the concepts in this chapter.

2. Saint Leo University is an educational enterprise. All of us, individually and collectively, work hard to ensure that our students develop the character, learn the skills and assimilate the knowledge essential to become morally responsible leaders. The success of our University depends upon a conscientious commitment to our mission, vision and goals. Saint Leo has used the phrase Points of Pride in reference to some of the results of this continuing commitment. A few of those points in the past have included:
   - One of the largest providers of higher education to the military

- Ranked second in the nation in the number of bachelor's degrees in business awarded to African-American students
- One of the largest providers of higher education online
- Ranked as the third largest among private universities and colleges in Florida
- The largest of three Catholic institutions in Florida
- Ranked eighth in overall enrollment among Catholic colleges and universities

Break into teams, choose any of the "Points of Pride" for Saint Leo University, and back up or refute the statement using any effective combination of statistical graphs and data analysis tools. Yes, you will have to use the web to do some research.

Values and Critical Thinking for Problem Solving:

1. Does it Add Up: What's Your Grade?
   A mean is an average. A weighted mean can occur when different values hold different weights in the final average. Choose sample scores for each one of your assignments in this class and calculate your final course grade using the idea of a weighted mean. Is a course grade something in which to take pride in? Can your effort in this course reflect the value of Excellence even though your grade may not?

2. What is a set of data that would have a skewed distribution and why is it skewed? Is there any bias as a result of the skewed nature of the data you found?

3. Why is it important that a sample be representative of the population it came from?

4. Use any news source to find an example of a study or graph that is deceptive and explain why you believe it is deceptive and what should have been done differently.

5. If a study showed that many Business majors drop out of school immediately after taking a Statistics class, what is wrong with the following statement? "A study indicates that the course requirement to take Statistics causes many Business majors to drop out of school." What are other possible lines of reasoning that might be true rather than Statistics causing people to drop out of school? What might be a correct interpretation of the study?

Writing across the Curriculum:

1. Compare and contrast mean, median, mode, and midrange. Be sure to explain the advantages and disadvantages of each.

2. Explain how a score of 68% on one test could possibly be considered better than a score of 85%

on a different test using the idea of relative standing.

3. Choose any topic, such as basketball. Give an example of something related to that topic that could be studied as a cross sectional study such as determining the rankings for highest free-throw percentages. Give an example of a retrospective study related to that topic such as how the number of concussions in basketball over the last 5 years compares to other sports. Finally, give an example of a prospective study on the same topic such as following a group of 100 students who played basketball in high school to compare their bone density in their 20s, 30s, 40s, and 50s to those who did not play any sports.

4. Write a short summary that describes pitfalls to be aware of when reading statistical studies. Be sure to include examples.

Social Justice across the Curriculum:

1. Why is the median price used to reflect average home prices in most areas, rather than the mean or mode? Should all people have similar houses or is it fair that some houses are much more expensive than others? Why or why not?

2. Some universities use SAT and ACT scores to determine whether to accept students. How can those two different scores be compared when they are on different scales and have completely different numeric values? Do you think test scores

are a fair way to determine entrance eligibility to a college or university? Why or why not?

3. If the median income for a certain area is far above poverty level, does that mean the area has no problem with poverty? Why or why not? What would have to be true about the dispersion in order to be sure that there is no poverty problem in this area? Are there any other statistics that are important to understand in order to answer this question?

For each of these questions, consider how the value of Respect relates to the situation and the importance of respect when conducting a statistical study.

1. Our national debt can be found in many different sources with the statistics represented many different ways. Identify three or more of these sources and discuss whether any of them provide a plan for the future generations to be able to pay for our spending.

2. What are your views on the country spending more money than it takes in? Support your views with statistical reports or documentation.

3. An "incentive" is often used to encourage people to participate in a study, such as free tickets to an event, money donated to a charity, a chance to win a prize, or even cash. How can an incentive create bias within the study? What group might likely not be included in a study and what group might be over represented when an incentive is used?

Data Analysis:

Choose any of the following questions to analyze. Your analysis should be submitted as an electronic file that contains the data you used, a citation for the source of the data, a summarized table such as a frequency distribution table and a relative frequency distribution table. At least two forms of graph or charts that visually summarize the data. At least three appropriate measures of central tendency such as mean, median, mode and/or midrange and at least two measures of dispersion such as standard deviation, variance, and/or range. Also include a minimum of one paragraph analysis of what your findings imply or tell you about the question you are analyzing. You may also break the paragraph into multiple parts like the examples in this chapter.

1. Were deaths of children in the U.S. ages 1-14 more of a problem in 1985 or 1991?
2. Was child poverty more of a problem in 1985 or 1991 in the U.S.?
3. Was teen suicide more of a problem in the U.S. in 1985 or 1991?
4. Was median income higher in 1987-1988 or 1988-1989?
5. Were juvenile crime arrest rates higher in 1984 or 1988? How about 1994?
6. Choose a sport and two positions, which gets paid more?
7. What is the population per square mile in the U.S.? (this answer should be a description of

highs, lows, averages, etc., possibly broken down by states, not a single value)

8. Is body fat percentage higher in older men?
9. Do taller men have a lower body fat percentage?
10. Do older men weigh more than younger men?
11. Do men with higher levels of body fat have larger wrists?
12. Are older men taller?
13. Do genders have the same body temperature?
14. Do infielders weigh more than outfielders? (or compare any other positions in the baseball almanac)
15. Are infielders taller than outfielders? (or compare any other two positions in the baseball almanac)
16. Are Infielders older than outfielders? (or compare any other positions in the baseball almanac)

Made in the USA
San Bernardino, CA
31 August 2015